Cambridge Elements ☰

Elements of Emerging Theories and Technologies
in Metamaterials
edited by
Tie Jun Cui
Southeast University, China
John B. Pendry
Imperial College London

SPOOF SURFACE PLASMON
METAMATERIALS

Paloma Arroyo Huidobro
Imperial College London

Antonio I. Fernández-Domínguez
*Autonomous University of Madrid and Condensed
Matter Physics Center (IFIMAC)*

John B. Pendry
Imperial College London

Luis Martín-Moreno
*Aragón Materials Science Institute (ICMA)
and University of Zaragoza*

Francisco J. García-Vidal
*Autonomous University of Madrid and Condensed
Matter Physics Center (IFIMAC)*

CAMBRIDGE
UNIVERSITY PRESS

University Printing House, Cambridge CB2 8BS, United Kingdom

One Liberty Plaza, 20th Floor, New York, NY 10006, USA

477 Williamstown Road, Port Melbourne, VIC 3207, Australia

314–321, 3rd Floor, Plot 3, Splendor Forum, Jasola District Centre,
New Delhi – 110025, India

79 Anson Road, #06–04/06, Singapore 079906

Cambridge University Press is part of the University of Cambridge.

It furthers the University's mission by disseminating knowledge in the pursuit of
education, learning and research at the highest international levels of excellence.

www.cambridge.org
Information on this title: www.cambridge.org/9781108451055
DOI: 10.1017/9781108553445

First published 2018

A catalogue record for this publication is available from the British Library.

ISBN 978-1-108-45105-5 Paperback
ISSN 2399-7486 (Online)
ISSN 2514-3875 (Print)

Spoof Surface Plasmon Metamaterials

Paloma Arroyo Huidobro, Antonio I. Fernández-Domínguez, John B. Pendry, Luis Martín-Moreno and Francisco J. García-Vidal

Abstract: *Metamaterials offer the possibility to control and manipulate electromagnetic radiation. These artificial materials, which are structured at a scale much smaller than the wavelength of radiation, show unusual and enhanced responses to electromagnetic waves otherwise not available from natural materials. An archetypical instance of metamaterials makes use of metallic nanostructures which benefit from the greatly enhanced light-matter interaction provided by the surface plasmons sustained by metals at high frequencies. While at lower frequencies metal surfaces do not present a plasmonic response, a high confinement of electromagnetic energy can still be achieved through the so-called spoof surface plasmons. These are surface electromagnetic waves sustained by periodically structured metal, and can be used to design plasmonic metamaterials at low frequencies. Their geometrical origin provides a great flexibility for tuning their properties.*

Spoof surface plasmon metamaterials are the focus of this Element of the Metamaterials Series. The fundamentals of spoof surface plasmons are reviewed, and advances on plasmonic metamaterials based on spoof plasmons are presented. Spoof surface plasmon metamaterials on a wide range of geometries are discussed: from planar platforms to waveguides and localized modes, including cylindrical structures, grooves, wedges, dominos or conformal surface plasmons in ultrathin platforms. The Element closes with a review of recent advances and applications such as Terahertz sensing or integrated devices and circuits.

1

Keywords: plasmonics, metamaterials, metasurfaces, spoof surface plasmons, terahertz waveguides, coupled mode method, subwavelength particles, magnetic resonance

ISSNs: 2399-7486 (Online), 2514-3875 (Print)

ISBNs: 9781108451055 PB, 9781108553445 OC

---

1 Introduction

The control and manipulation of light constitutes a major aim for current technology. For more than a decade, metamaterials have offered the possibility of governing electromagnetic (EM) fields in unprecedented ways [1, 2]. Metamaterials are artificial materials whose optical properties are determined through their bulk or surface structuring at a deeply subwavelength scale. These devices have allowed physicists, nanoscientists and engineers to realize extraordinary EM phenomena not found in nature [3, 4]. For instance, while conventional materials usually respond mostly to electric fields, metamaterials can be designed to feature a strong magnetic response too, which greatly broadens the range of applications enabled by metamaterial technology [5, 6].

Noble metals are an archetypical example of conventional materials presenting a remarkably strong electric response. The optical properties of metals are assisted by surface plasmons [7]. These are greatly localized light-matter excitations, originating from the interaction between EM fields and free electrons in metals, and have the striking ability of concentrating EM energy efficiently at the nanoscale [8]. The unique properties of surface plasmons have been widely employed in chemical and bio-sensing, energy harvesting and super-resolution imaging [9–11].

Plasmonic metamaterials combine the intrinsic strong electric response of metals with the versatility and tunability of metamaterial designs in order to realize novel light phenomena. A paradigmatic example of these devices is the so-called double fishnet metamaterial, which presents an effective negative refractive index in the optical regime [12, 13]. In the metamaterial context, surface plasmons have allowed the implementation at high frequencies of effects firstly reported for radio, microwave or terahertz (THz) waves. However, the fact that metal surfaces sustain these confined EM modes only at visible wavelenghts prevents the transferring of plasmonic capabilities to lower frequency domains. Nonetheless, structured metal surfaces can mimic plasmonic abilities (such as high confinement and amplification) at low frequency ranges of the EM spectrum, as they support spoof surface plasmon modes even when described as perfect conductors [14, 15]. Importantly, unlike their optical counterparts, spoof surface plasmons have a purely geometric origin.

This Element reviews the fundamentals of spoof surface plasmons, as well as recent advances in the new kind of plasmonic metamaterials based on spoof plasmons. As a general introduction, this Section firstly summarizes key concepts in the fields of metamaterials (Section 1.1) and plasmonics (Section 1.2). Then, in Section 1.3, a brief overview of spoof surface plasmons is provided, together with the outline of this Element.

1.1 Metamaterials

A new class of EM materials has offered a way to overcome the limitations of natural materials. Metamaterials are artificial materials whose EM properties are given not by their atomic or chemical composition but rather by their artificial structure [1–4]. They are made of engineered building blocks that are smaller than the wavelength of free space radiation under consideration. Owing to such subwavelength constituents, which act as artificial "atoms" or "molecules" that respond in a prescribed way to the EM fields, metamaterials can be designed to have properties that do not occur in natural materials [5, 6, 16–18].

The two basic building blocks employed in the design of metamaterials are shown in Figure 1.1: metallic wires [19], depicted in panel (a), and split-ring structures [20, 21], shown in (b). Both wires and rings support resonances that yield different and tunable responses to EM fields. While metallic wires allow for low-frequency plasmons by reducing the electron density, metallic

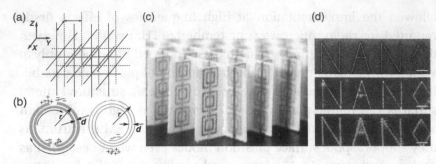

Figure 1.1 Metamaterials are artificially structured materials whose properties are determined by their subwavelength constituents, such as wires and rings. (a) Metallic wires arranged in a 3D lattice behave as an isotropic low frequency plasma [19]. (b) Split-ring resonators (left) display a magnetic moment when a magnetic field induces currents along the metal rings [20]. Swiss rolls (right) act similarly, being resonant at lower frequencies due to their higher capacitance [21]. A combination of such basic structures results in effective electric and magnetic responses. For instance, a material with a negative index of refraction at microwave frequencies was realized by means of copper split-ring resonators (c) arranged in a unit cell of 5 mm. (d) Experimental demonstration of a silver superlens that focuses propagating and evanescent waves owing to its negative index of refraction. (a–b) Reproduced from [45]. (c) Reprinted from [3] and (d) from [26].

split rings produce artificial magnetism. An arrangement of these two basic structures acts as a homogeneous medium for EM fields as long as their sizes and spacing are much smaller than the wavelength of the EM waves. In the effective medium approximation, the collective response of the subwavelength constituents effectively yields a wide range of values of the electric permittivity, $\hat{\epsilon}$, and the magnetic permeability, $\hat{\mu}$. Then, a prescribed EM response can be obtained by an appropriate design of the metamaterial. For instance, while magnetism can only be found in nature at low frequencies below the THz regime, negative index materials with simultaneously negative electric permittivity and magnetic permeability have been realized from the optical to the GHz ranges of the spectrum; see Figure 1.1 (c) [3, 22–25]. Further advances that have been enabled by metamaterials include sub-diffraction limited resolution [26–28] – see Figure 1.1 (d); chirality [29, 30] and toroidal dipole moments [31] for different applications; and high resolution biosensing [32–34]. In addition, the development of three-dimensional (3D) [35] and nonlinear and active metamaterials [36, 37] has also been pursued.

More recently, considerable attention has been devoted to meta-surfaces, the 2D counterpart of metamaterials [38]. Consisting of a planar arrangement of resonant subwavelength-sized building blocks, metasurfaces allow for planar photonic applications [39]. By appropriately designing the blocks and their arrangement, metasurfaces provide an ultrathin platform for manipulating electromagnetic (EM) waves. Novel phenomena based on metasurfaces range from broadband light bending and anomalous reflection and refraction [40–42] to strong spin–orbit interactions of light [43], and applications include planar metalenses with sub-wavelength resolution [44, 233].

1.2 Plasmonics

Surface plasmons (SPs) arise from the interaction between EM fields and the conduction electrons in a metal [7]. Conduction electrons can be driven by an EM field, which causes them to oscillate collectively. These charge density oscillations can be sustained within the bulk of a metal as well as confined at the interface between a metal and a dielectric medium, in each case at a distinct resonance frequency. Spatially confined plasmon modes at metal/dielectric interfaces, i.e., SPs, are the subject of study of the field of plasmonics [46]. The fact that SPs are collective excitations of photons and electrons gives rise to a major characteristic: the subwavelength nature of SPs enables a strong localization and enhancement of the EM fields in the vicinity of metal/dielectric interfaces.

As we describe below, two kinds of SPs exist, depending on the geometry of the interface under consideration. First, flat metal surfaces support the propagation of *surface plasmon polaritons* (SPPs), EM modes that propagate along the surface while featuring a subwavelength confinement to it. Many research efforts have been devoted to the implementation of SPP-based photonics [9], as metal structures can be designed to efficiently guide SPPs by, for instance, confining them in the transverse direction in one-dimensional (1D) plasmonic waveguides [11, 47]. Moreover, SPPs offer a bridge between photonics and electronics, opening up the possibility of achieving more compact devices [10]. On the other hand, metal particles of dimensions smaller than the wavelength sustain confined SP modes at optical frequencies, the so-called *localized surface plasmons* (LSPs). These are collective oscillations of the conduction electrons that give rise to subwavelength localization of the EM fields. LSPs in metal nanoparticles enable an efficient transfer of energy between the near and far fields, and for this reason metal nanoparticles are considered as nanoantennas, the analog of radio antennas but at higher frequencies [48–61].

The existence of SPs (both SPPs and LSPs) is a characteristic feature of the EM response of metals. The optical response of a material can be characterized by means of the dielectric permittivity, $\epsilon(\omega)$, which for metals depends strongly on the frequency of light, ω. The Drude theory considers a metal as a background of fixed core ions plus a free electron gas in order to appropriately model conduction electrons, which are nearly free to move within the bulk. Within this model the electric permittivity can be written as

$$\epsilon(\omega) = \epsilon_\infty \left(1 - \frac{\omega_p^2}{\omega^2 + i\gamma\omega} \right), \tag{1.1}$$

where ω_p is the plasma frequency, γ is the damping constant of electrons and ϵ_∞ is a high frequency offset. The bulk plasma frequency is characteristic of each metal and reads as

$$\omega_p = \sqrt{\frac{ne^2}{m_e\epsilon_0}}, \qquad (1.2)$$

with n being the electron density, e and m_e the electron charge and effective mass, and ϵ_0 the vacuum permittivity. For most metals, ω_p lies in the ultraviolet (UV) regime, $\omega_p \sim 10^{16}$ Hz (or, equivalently, a few eV). The real and imaginary parts of the electric permittivity for silver are plotted in Figure 1.2 (panels (a) and (b), respectively). The results obtained with the Drude model (Equation 1.1) for parameters $\hbar\omega_p = 4.2$ eV, $\hbar\gamma = 0.07$ eV and $\epsilon_\infty = 4.6$ [62], are plotted as a solid black line, while the dashed lines correspond to two sets of experimental data (Johnson and Christy [63] and Palik [64]). Below the plasma frequency, $\omega < \omega_p$, the real part of $\epsilon(\omega)$ is negative, a fact that is well reproduced by the Drude model (see inset in panel (a)), and its imaginary part takes a nonzero value, reflecting the relaxation time of electrons in the lattice. On the other hand, the Drude model fails to give an accurate description

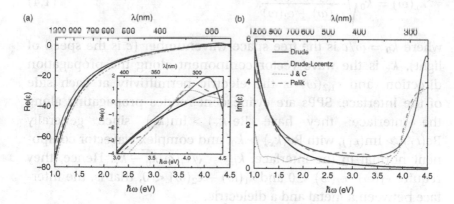

Figure 1.2 Optical response of Silver. Real (a) and imaginary (b) parts of the electric permittivity, $\epsilon(\omega)$, for silver in the optical regime. Two sets of experimental data are plotted (Johnson and Christy [63] and Palik [64]) together with the Drude model, given by Equation 1.1, and the Drude–Lorentz model. Note that other noble metals, such as gold, feature similar properties.

at frequencies $\omega \gtrsim \omega_p$, where the imaginary part of $\epsilon(\omega)$ increases
due to interband transitions – since at high frequencies photons can
promote bound electrons to the conduction band. As shown in Figure
1.2, a more accurate modeling of $\epsilon(\omega)$ for higher frequencies is given
by the Drude–Lorentz model (see, for instance, [65] for more details).

Let us now focus our attention on the propagating SP modes.
SPPs are p-polarized (magnetic field component lying in the plane
of the interface) and their field intensity decays exponentially in the
direction perpendicular to the interface. A sketch of a SPP propagating through the interface separating a metal and a dielectric
(permittivities $\epsilon(\omega)$ and ϵ_d, respectively) is shown in Figure 1.3.
SPPs are solutions of Maxwell's equations bound to an interface,

$$\nabla \times \nabla \times \mathbf{E}(\mathbf{r},\omega) - \frac{\omega^2}{c^2} \epsilon(\mathbf{r},\omega) \mathbf{E}(\mathbf{r},\omega) = 0. \tag{1.3}$$

Their dispersion relation can be written as

$$k_x(\omega) = k_0 \sqrt{\frac{\epsilon_1(\omega)\epsilon_2(\omega)}{\epsilon_1(\omega) + \epsilon_2(\omega)}}, \tag{1.4}$$

where $k_0 = \omega/c$ is the free space wavenumber (c is the speed of
light), k_x is the wave vector component along the propagation
direction, and $\epsilon_{1,2}(\omega)$ is the electric permittivity at each side
of the interface. SPPs are evanescent modes propagating along
the interface: they have $\mathrm{Re}(k_x) > \mathrm{Im}(k_x)$ since generally
$\mathrm{Re}(\epsilon_1) \gg \mathrm{Im}(\epsilon_1)$, with $\mathrm{Re}(k_x) > k_0$, and complex wavevector component normal to the interface, $k_{j,z} = \sqrt{\epsilon_j(\omega)k_0^2 - k_x^2}$. Hence, they
require $\epsilon_1(\omega) \cdot \epsilon_2(\omega) < 0$ and $\epsilon_1(\omega) + \epsilon_2(\omega) < 0$, that is, the interface between a metal and a dielectric.

The dispersion relation given by Equation 1.4 is plotted in
Figure 1.3 for a silver/air interface. We take $\epsilon_2 = 1$ and $\epsilon_1(\omega)$
from the Drude model for silver (Equation 1.1). According to
Equation 1.4, and assuming a lossless metal, $\gamma = 0$, frequencies
below $\omega_{sp} = \omega_p/\sqrt{\epsilon_\infty^{-1} + 1}$ yield $k_x > k_0$, i.e., the mode lies below
the light line ($\omega = ck_x$, plotted as yellow line in the figure). This is

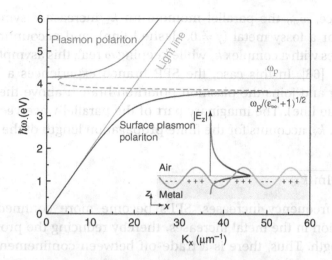

Figure 1.3 SPPs propagating along a metal/dielectric (air) interface. The plot shows the dispersion relation (energy versus parallel momentum) for the EM modes supported by a silver/air interface. We take the Drude model for the permittivity of silver, with the same parameters as in Figure 1.2, and plot two cases: a lossless ($\gamma = 0$, solid lines) and a lossy metal (dashed line). The light line, given by $\omega = ck_x$, is also given. The SPP branch lies below the light line and approaches the SP resonance frequency, $\omega_{sp} = \omega_p/\sqrt{\epsilon_\infty^{-1} + 1}$. The plasmon polariton branch sets off at the bulk plasmon resonance frequency, ω_p, approaches the light line from above as the frequency increases. Inset panel: SPPs have a mixed nature between light and charges.

the SPP branch (see the red line). At low frequencies, $\omega \ll \omega_{sp}$, $|\epsilon(\omega)| \gg 1$ and $k_x \approx k_0$, meaning that the SPP is poorly confined to the metal. As the frequency increases, approaching the optical regime, $|\epsilon(\omega)|$ decreases and $k_x \gtrsim k_0$, leading to an increase in the binding of the mode to the interface: the confinement of the SPP in the perpendicular direction reaches subwavelength values. Hence, the characteristics of the optical response of metals are responsible for the existence of SPPs, which are not tightly confined at lower frequency ranges. Finally, as the frequency approaches the SP resonance, ω_{sp}, the parallel momentum k_x increases asymptotically. For a lossy metal ($\gamma \neq 0$, dashed line), and accounting for the losses with a complex k_x while keeping ω real, this asymptote is avoided [66]. In this case, the SPP branch experiences a back-bending and joins the plasmon polariton branch above the light line (blue line). The imaginary part of the parallel wave vector of the SPP, k_x, accounts for the finite propagation length of the SPP,

$$L_p = \frac{1}{\mathrm{Im}(k_x)}. \tag{1.5}$$

As the frequency increases, SPPs become more confined and absorption in the metal increases, thereby reducing the propagation length. Thus, there is a trade-off between confinement and propagation [7].

On the other hand, the fact that SPPs are evanescent modes, for which $\mathrm{Re}(k_x) > k_0$, means that they cannot be excited directly with free space photons. Instead, the momentum mismatch has to be overcome by, for instance, patterning a grating in the metal surface. On the other hand, in the Otto [67] or Kretschmann [68] configurations the coupling mechanism is provided by another dielectric medium of refractive index n, that lowers the light line to $\omega = ck_x/n$, thus crossing the SPP dispersion curve. Alternatively, near-field probes or quantum emitters, such as fluorescent molecules, can also be used to excite SPPs when placed in the close proximity of the interface, as they couple to all available parallel momenta.

A different kind of SPs arises in metal particles with closed surfaces [69]. When light impinges on a metallic particle of dimensions much smaller than the wavelength of radiation, the electron gas in the metal gets polarized, with polarization charges localized at the surface. As a consequence, a restoring force emerges and originates a plasmon oscillation that is confined at the surface of the metal nanoparticle. This resonance is the LSP [8], and, differently from SPPs, can be directly excited by light. An insightful model for understanding the LSP resonance is that of a Rayleigh spherical particle (dimensions $\ll$ wavelength of the incident light). In the quasi-static limit, the optical response of a subwavelength particle is determined by its polarizability,

$$\alpha = 4\pi R^3 \frac{\epsilon_m(\omega) - \epsilon_d}{\epsilon_m(\omega) + 2\epsilon_d}, \qquad (1.6)$$

with R being the radius of the particle and ϵ_d being the electric permittivity of the surrounding medium. From the above expression, it is clear that the polarizability is resonant under the condition $\epsilon_m(\omega) + 2\epsilon_d = 0$, which, for a lossless Drude model, results in the following resonance frequency:

$$\omega_{LSP} = \frac{\omega_p}{\sqrt{\epsilon_\infty + 2\epsilon_d}}. \qquad (1.7)$$

The resonance in the polarizability implies an enhancement of the dipolar EM field generated by the metal particle [see Figure 1.4]. In general, the resonance frequency of LSPs is determined by the

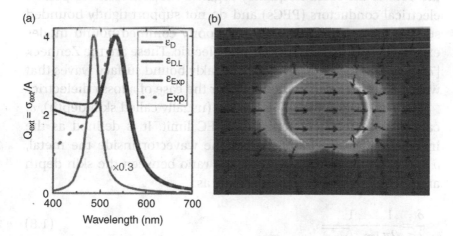

Figure 1.4 LSPs and optical antennas. (a) Extinction cross section for a 25 nm radius gold sphere in water. (b) EM field scattered by the LSP resonance in the metal nanoparticle upon illumination at $\lambda = 535$ nm.
Reprinted from [70].

shape, size and material properties of the metal nanoparticle. The ability of LSPs to confine light in a subwavelength scale is accompanied with a great enhancement of the EM energy, which allows for the localization of light even beyond the diffraction limit. Thus, metal nanoparticles have found applications in biomedical sensing, surface enhanced spectroscopies, near-field microscopy or photovoltaics [70].

As we have discussed, the properties of SPs are closely related to the behavior of metal permittivity close to the plasma frequency. For this reason, plasmonic effects are in principle limited to the high frequency range of the spectrum (UV, optical frequencies and near IR). At low frequencies ($\omega \ll \omega_{\mathrm{p}}, \gamma$), the permittivity of metals behaves very differently, with significantly large values of its (negative) real and imaginary parts. In this regime, metal permittivity is dominated by the conductivity, $\epsilon(\omega) \sim i\sigma/\omega$, and free electrons respond immediately to any EM perturbation, thereby shielding the metal from the EM field [71]. As a consequence, from the THz to the radio frequency regimes metals resemble perfect electrical conductors (PECs) and do not support tightly bounded surface EM modes. Instead, only poorly confined bound modes exist, which extend mostly in the dielectric. These are the Zenneck [72] and Sommerfeld [73] waves, weakly bound surface waves that were first introduced by Zenneck for the case of a lossy dielectric.

The penetration depth of a metal (usually called skin depth), δ, can be used to characterize the PEC limit. It is defined as the inverse of the imaginary part of the wavevector inside the metal, $\delta = 1/\mathrm{Im}\, k$. Since $k = k_0 \sqrt{\epsilon(\omega)}$, the ratio between the skin depth and the operating wavelength reads as:

$$\frac{\delta}{\lambda} = \frac{1}{2\pi} \frac{1}{\mathrm{Im}\sqrt{\epsilon(\omega)}}. \tag{1.8}$$

This normalized magnitude (and not δ) is closely related to the impedance of a metal surface, $Z_S(\omega) = 1/\sqrt{\epsilon(\omega)}$, which measures how much the parallel component of the electric field at a metal surface, $\vec{E}_{\parallel}$, departs from its PEC value, $\vec{E}_{\parallel} = 0$. For instance, δ/λ is of the order of 10^{-5} for silver at GHz frequencies and increases as the frequency increases. Therefore, PEC approximation (that assumes surface impedance $Z_S = 0$) is very accurate for very low frequencies and becomes worse for higher frequencies.

Finally, it should be noted that semiconductors can support SPs at lower frequencies than metals. Due to their lower free-carrier density their plasma frequencies are lower than in metals, falling in the THz regime [74, 75].

1.3 Spoof Surface Plasmons: Designing Plasmonic Metamaterials

While metals do not support tightly bound SPs at the lower frequency ranges of the spectrum, such as the microwave or THz regimes, it would be highly desirable to transfer the SP ability to localize light in deep subwavelength dimensions from the visible and near infrared (IR) range to lower frequencies. Since the middle of the last century, it has been known that the addition of a subwavelength corrugation (arrays of holes, for example) to a metal surface produces an enhanced surface impedance. This mechanism allows a surface mode to be bound to the interface, even in the limit of perfect conductivity [76, 77], and is at the origin of the broad field of frequency-selective surfaces [78]. This effect can be understood by considering the presence of a periodic array of small holes within a perturbative approach. Thus, the dispersion relation of the surface EM modes sustained by the corrugated structure will closely follow the one for the flat SPPs (or Zenneck waves), except for values of the momentum lying close to the boundary of the Brillouin Zone. There, the band of the corrugated surface bends in order to accommodate for band gaps. As a consequence of the band bending caused by the periodic array of holes, the lowest band of the surface EM modes separates from the light line, binding the EM field more strongly to the surface [79]. In particular, the dispersion relation below the first gap represents a truly bound surface mode.

In 2004, Pendry, Martín-Moreno and García-Vidal gave an additional insight to the understanding of periodicity-induced binding of

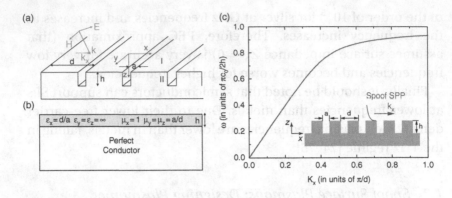

Figure 1.5 Spoof SPPs propagating along a periodic array of grooves. (a) A 1D array of grooves of width a, period d and depth h. (b) In the effective medium approach, the array of grooves behaves as a layer with homogeneous but anisotropic EM properties, on top of a perfect conductor. (c) Dispersion relation of the spoof SPPs supported by the structure shown in (a). The geometrical parameters of the grooves are $a/d = 0.2$ and $h/d = 1$.
The figure has been adapted from [15].

the EM fields to the surface [14]. These authors found that the surface corrugation can be effectively described as a surface layer with a dielectric permittivity of the Drude form, whose plasma frequency is given by the cutoff frequency of the hole waveguide. In this way, the surface EM modes supported by corrugated metal surfaces can be entirely controlled by geometry, in a similar fashion to the already introduced metamaterials [15]. In the PEC limit, these designed surface EM modes are known as *spoof* surface plasmon polaritons. Figure 1.5 illustrates the spoof SPPs sustained by a periodic array of 1D grooves carved on a PEC surface. Panel (a) sketches the actual supporting geometry, and panel (b) shows the corresponding effective medium model. Panel (c) plots the spoof plasmon dispersion relation for this system below the first band gap.

The proposal of spoof SPPs in 2D structured flat surfaces was followed by subsequent experimental demonstrations in the microwave [80, 81] and THz regimes [82, 83], as well as by upgrades of the original theoretical model [84, 85]. In addition, its extension to laterally confined geometries, which opens up the possibility of designing waveguides, has been extensively studied both in theoretical and experimental works. Theoretical considerations include periodically corrugated wires [86], wedges [87], channels [88], periodic chains of dominoes [89] or metal–insulator–metal waveguides [90]. On the other hand, among the experimental realizations of waveguides for spoof SPPs we find helically grooved wires [91] and the experimental verification of the so-called domino plasmons [92]. In the THz and microwave regimes, where most work has been focused, spoof SPPs display deep subwavelength confinement as well as long enough propagation lengths. Regarding higher frequencies, spoof SPPs have also been applied for tailoring the guiding properties of corrugated metal waveguides working at optical and telecom frequencies [93]. In addition, the concept of spoof SPPs has also enabled the interesting prospect of propagation along curved surfaces. As introduced in [94], conformal surface plasmons, a kind of spoof SPPs sustained by ultrathin films, are able to propagate along arbitrarily curved surfaces with very low propagation losses. Finally, the concept of spoof SPPs has also been extended to localized geometries. Corrugated metal particles support the analogues of plasmonic resonances but at lower frequencies [95]. This allows the transfer of all the capabilities of plasmonics to frequencies away from the optical regime, such as subwavelength dipolar resonances and huge field enhancements in the gaps between particles, with the addition of modes of magnetic character [96].

This Element focuses on the progress in the field of surface EM waves on structured metal surfaces reported since 2004. A modal expansion formalism specifically developed for the description of spoof SPP phenomena is presented in Section 2. In the following sections, the application of the spoof SPP concept to a wide range of geometries is discussed. Planar platforms are first considered in Section 3. Spoof surface plasmon waveguides in cylindrical geometries and more complex structures, such as V-grooves, wedges, domino or conformal spoof plasmons, are reviewed in Section 4. Finally, localized spoof surface plasmons, the analog to the optical

LSP resonances supported by metal nanoparticles, are presented in Section 5.

2 Theoretical Formalism: Coupled Mode Method

We introduce the theoretical formalism that we employ to describe the geometry dependence of the modal properties of the spoof SPP supported by planar metal structures partially or fully perforated with periodic arrays of indentations. This framework will be extended to spoof SPP cylindrical waveguides and spoof LSP resonators in the following Sections. It is based on the modal expansion technique, a powerful theoretical framework which has been successfully applied to study different EM phenomena such as extraordinary transmission [97, 98] or negative index metamaterials [99]. As a difference with other approaches used to analyze the emergence of spoof SPP modes [84, 85, 100], this theoretical method yields, under certain conditions, analytical expressions for the dispersion relation of the guided modes. Despite their approximate character, these simple dispersion relations allow us to reach a deeper understanding of the spoof SPP concept and their potential applications.

The approach consists in the expansion of the EM fields into eigenmodes of Maxwell's equations within the various regions comprising the system under study. By imposing the appropriate boundary conditions at all the interfaces, EM fields can be constructed in all space. Although the so-called surface impedance boundary conditions [71] can be implemented into the formalism, in this Section we treat metals as PECs, which is an excellent approximation at microwave or THz frequencies. Note that the scattering properties of a corrugated PEC remain invariant if all the lengths are scaled by the same factor, which allows the transfer of results from one frequency range to another.

Figure 2.1 shows a schematic picture of the expansion procedure for the case of a 2D array of square holes with periods d_x and d_y. Here, we illustrate the construction of our theoretical framework in

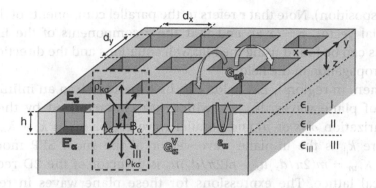

Figure 2.1 Schematic picture of the modal expansion procedure for the case of a periodic array of square holes. Both the expansion coefficients and the different terms in Equations 2.12 and 2.16 are rendered schematically. Reproduced from [101].

planar structures. As mentioned above, in Sections 4.1 and 5.1 we will introduce the modifications which allow us to treat cylindrical and localized geometries, respectively. We denote by z the direction normal to the metal structure. The interfaces are placed at $z = 0$ and $z = h$. When treating fully pierced structures, the latter interface corresponds to the lower surface of the metal slab, whereas in the case of partial perforation, it gives the bottom of the blind indentations decorating the upper metal surface. Thus, the system is divided into three regions along the z-direction: the upper semi-space with dielectric constant ϵ_I (region I), the metal perforated with filled indentations with ϵ_{II} (region II), and the substrate, comprising a dielectric with permittivity ϵ_{III} (a PEC medium) for fully (partially) pierced geometries (region III). Note that the periodic character of the system enables us to apply Bloch's theorem and expand the EM fields only within the unit cell of area $d_x \times d_y$ containing one single indentation.

A convenient notation, which simplifies the calculations presented in this section, is to use a Dirac nomenclature for the EM fields. In this way, we define the bi-vectors $< \mathbf{r}|\mathbf{E} >= \mathbf{E}(\mathbf{r}) = (E_x(\mathbf{r}), E_y(\mathbf{r}))^T$ and $< \mathbf{r}|\mathbf{H} >= \mathbf{H}(\mathbf{r}) = (H_x, H_y)^T$ (T standing for transposition). Note that **r** refers to the parallel components of the spatial vector, $\mathbf{r} = (x,y)$, and that the z-components of the EM fields can be found using the Maxwell equations and the direction of propagation of the field.

Then, in region I, EM fields can be expanded into an infinite set of plane waves, $|k_{mn},\sigma\rangle$, which are characterized by their polarization σ (s or p) and parallel wave vector $k_{mn} = k_{\|} + K_{mn}$, where $k_{\|}$ is the in-plane wave vector of the spoof SPP mode and $K_{mn} = m(2\pi/d_x)u_x + n(2\pi/d_y)u_y$ is a vector of the 2D reciprocal lattice. The expressions for these plane waves in real space are:

$$< \mathbf{r}|\mathbf{k},p > = (k_x,k_y)^T \exp(i\mathbf{k}\cdot\mathbf{r})/\sqrt{L_xL_y|\mathbf{k}|^2} \tag{2.1}$$

$$< \mathbf{r}|\mathbf{k},s > = (-k_y,k_x)^T \exp(i\mathbf{k}\cdot\mathbf{r})/\sqrt{L_xL_y|\mathbf{k}|^2}.$$

These modes are orthonormal when integrated over a unit cell, i.e., $< \mathbf{k},\sigma|\mathbf{k}',\sigma' > = \delta_{\mathbf{k},\mathbf{k}'}\delta_{\sigma,\sigma'}$, where δ is the Kronecker delta.

By introducing the unknown expansion coefficients, $\rho^I_{k_{mn}\sigma}$, the parallel components of the electric and magnetic fields can be written as

$$|E^I_t\rangle = \sum_{m,n,\sigma} \rho^I_{k_{mn}\sigma}|k_{mn},\sigma\rangle e^{\kappa^I_{mn}z}, \tag{2.2}$$

$$|-u_z \times H^I_t\rangle = \sum_{m,n,\sigma} Y^I_{k_{mn}\sigma}\rho^I_{k_{mn}\sigma}|k_{mn},\sigma\rangle e^{\kappa^I_{mn}z}, \tag{2.3}$$

where the normal plane wave vector is $\kappa^I_{mn} = \sqrt{|k_{mn}|^2 - \epsilon_I k_0^2}$, and $k_0 = 2\pi/\lambda = 2\pi f/c$ is the wave vector modulus in vacuum. The electric and magnetic fields are related through the modal admittances $Y^I_{k_{mn}s} = i\kappa^I_{mn}/k_0$ and $Y^I_{k_{mn}p} = -i\epsilon_I k_0/\kappa^I_{mn}$. Note that we have assumed that all plane waves in region I are evanescent along the z direction, i.e., $|k_{mn}| > \epsilon_I k_0$.

In region II, as we are modeling the metal response through the PEC approach, EM fields are nonzero only within the perforations.

Therefore, parallel components of the fields can be expressed in terms of the corresponding waveguide modes, labeled with index α, having

$$|E_t^{II}\rangle = \sum_{\alpha} [A_\alpha e^{iq_\alpha(z-h)} + B_\alpha e^{-iq_\alpha(z-h)}|\alpha\rangle, \qquad (2.4)$$

$$|-u_z \times H_t^{II}\rangle = \sum_{\alpha} Y_\alpha^{II} [A_\alpha e^{iq_\alpha(z-h)} - B_\alpha e^{-iq_\alpha(z-h)}|\alpha\rangle, \qquad (2.5)$$

where q_α is the propagation constant of mode $|\alpha\rangle$ and A_α and B_α are the unknown expansion coefficients. The mode admittances are given by $Y_\alpha^{II} = q_\alpha/k_0$ (for s-polarization) and $Y_\alpha^{II} = \epsilon_{II} k_0/q_\alpha$ (for p-polarization). Note that in this case we do not impose the propagating/evanescent character of the basis elements along the z-direction. In the case of blind indentations, the PEC boundary at the bottom of the perforations implies $E_t|_{z=h} = 0$ and therefore $A_\alpha = -B_\alpha$ in Equations 2.4 and 2.5.

In region III, EM fields can be expressed again in terms of plane waves decaying along z-direction as

$$|E_t^{III}\rangle = \sum_{m,n,\sigma} \rho_{k_{mn}\sigma}^{III}|k_{mn},\sigma\rangle e^{-\kappa_{mn}^{III} z}, \qquad (2.6)$$

$$|-u_z \times H_t^{III}\rangle = -\sum_{m,n,\sigma} Y_{k_{mn}\sigma}^{III} \rho_{k_{mn}\sigma}^{III}|k_{mn},\sigma\rangle e^{-\kappa_{mn}^{III} z}, \qquad (2.7)$$

where the definitions of all terms are the same as in Equations. 2.2 and 2.3, substituting ϵ_I by ϵ_{III}. Importantly, when considering blind perforations, this region is filled with PEC metal and, therefore, electric and magnetic fields must vanish within it, i.e., $\rho_{k_{mn}\sigma}^{III} = 0$ for all $|k_{mn},\sigma\rangle$.

The unknowns $\rho_{k_{mn}\sigma}^{I}$, A_α, B_α and $\rho_{k_{mn}\sigma}^{III}$ are calculated by imposing continuity of the parallel components of the EM field at $z = 0$ and $z = h$. Thus, we obtain four vectorial equations (one for each field component at each interface) which depend on the parallel spatial coordinates x and y

$$\sum_{m,n,\sigma} \rho^{\mathrm{I}}_{\mathrm{k}_{mn}\sigma} |\mathrm{k}_{mn},\sigma\rangle = \sum_\alpha [A_\alpha e^{-iq_\alpha h} + B_\alpha e^{iq_\alpha h}]|\alpha\rangle, \qquad (2.8)$$

$$\sum_{m,n,\sigma} Y^{\mathrm{I}}_{\mathrm{k}_{mn}\sigma} \rho^{\mathrm{I}}_{\mathrm{k}_{mn}\sigma} |\mathrm{k}_{mn},\sigma\rangle = \sum_\alpha Y^{\mathrm{II}}_\alpha [A_\alpha e^{-iq_\alpha h} - B_\alpha e^{iq_\alpha h}]|\alpha\rangle, \qquad (2.9)$$

$$\sum_\alpha [A_\alpha + B_\alpha]|\alpha\rangle = \sum_{m,n,\sigma} \rho^{\mathrm{III}}_{\mathrm{k}_{mn}\sigma} |\mathrm{k}_{mn},\sigma\rangle e^{-\kappa^{\mathrm{III}}_{mn}h}, \qquad (2.10)$$

$$\sum_\alpha Y^{\mathrm{II}}_\alpha [A_\alpha - B_\alpha]|\alpha\rangle = -\sum_{m,n,\sigma} Y^{\mathrm{III}}_{\mathrm{k}_{mn}\sigma} \rho^{\mathrm{III}}_{\mathrm{k}_{mn}\sigma} |\mathrm{k}_{mn},\sigma\rangle e^{-\kappa^{\mathrm{III}}_{mn}h}. \qquad (2.11)$$

Note that in the case of blind indentations, Equations 2.10 and 2.11, which reflect field continuity at the boundary between regions II and III, are redundant. Let us stress that PEC approximation leads to a relevant difference between equations for parallel electric (Equations 2.8 and 2.10) and magnetic fields (Equations 2.9 and 2.11): whereas $\mathbf{E}_t$ must be continuous everywhere within the xy plane, $\mathbf{H}_t$ must be continuous only at the perforation openings. By projecting the electric continuity equations onto vacuum plane waves, and the magnetic ones onto indentation waveguide modes, we take into account this fact, and remove the dependence of the continuity equations on the spatial coordinates x and y. The analytical expressions for the overlap integrals between plane waves and waveguide modes of different aperture shapes can be found in the literature [102, 103].

We can combine the projected equations obtained from Equations 2.8–2.11 to express the continuity of EM fields at the two interfaces of the system in the form of tight-binding-like equations

$$(G^{\mathrm{I}}_{\alpha\alpha} - \Sigma_\alpha)E_\alpha + \sum_{\alpha' \neq \alpha} G^{\mathrm{I}}_{\alpha\alpha'} E_{\alpha'} - G^V_\alpha E'_\alpha = 0,$$

$$(G^{\mathrm{III}}_{\alpha\alpha} - \Sigma_\alpha)E'_\alpha + \sum_{\alpha' \neq \alpha} G^{\mathrm{III}}_{\alpha\alpha'} E'_{\alpha'} - G^V_\alpha E_\alpha = 0, \qquad (2.12)$$

whose unknowns are now the modal amplitudes of $\mathbf{E}_t$ at the openings of the perforations, which can be written in terms of the expansion coefficients as $E_\alpha = A_\alpha e^{-iq_\alpha h} + B_\alpha e^{iq_\alpha h}$ (at $z = 0$) and $E'_\alpha = -[A_\alpha + B_\alpha]$ (at $z = h$). We can give a simple physical interpretation to all the magnitudes appearing in the matching equations. First, let us stress that the upper equation in system 2.12 is obtained from Equations 2.8 and 2.9, and therefore it can be associated with the field continuity at the I–II interface of the system. Similarly, the lower equation is obtained from Equations 2.10 and 2.11, and is associated with continuity at the II–III interface. Thus, the term

$$\sum_\alpha = Y_\alpha^{II} \cot(q_\alpha h), \tag{2.13}$$

which takes into account how fields on one of the system interfaces affect the modal amplitudes at that interface, can be interpreted as resulting from the bouncing back and forth of the EM fields inside the indentations. On the other hand,

$$G_\alpha^V = Y_\alpha^{II} \frac{1}{\sin(q_\alpha h)}, \tag{2.14}$$

which links the amplitudes at one interface with the fields at the other, is reflecting the EM coupling at the two sides of the metal slab through the perforations. Note that these two terms involve only one waveguide mode $|\alpha\rangle$. Finally, the terms

$$G_{\alpha\alpha'}^{I,III} = i \sum_{m,n,\sigma} Y_{k_{mn}\sigma}^{I,III} \langle \alpha | k_{mn}\sigma \rangle \langle k_{mn}\sigma | \alpha' \rangle \tag{2.15}$$

describe the interaction between modal amplitudes corresponding to different waveguide modes at a given interface. Note that $Y_{k_{mn}p}^{I,III}$, and therefore the G-terms diverge when the grazing condition $\kappa_{mn}^{I,III} = \sqrt{|k_{mn}| - \epsilon_{I,III} k_0^2} = 0$ is satisfied. In Figure 2.1 the physical interpretation of the different terms in the set of homogeneous equations (Equations 2.12) are represented schematically.

Let us now describe how this result is modified for the case of blind holes, i.e., when a PEC substrate is considered. Then, Equations 2.8 and 2.9, together with the condition $A_\alpha = -B_\alpha$,

express the appropriate continuity of the EM fields. Following the
same matching procedure as before, we end up with a set of
equations of the form

$$(G_{\alpha\alpha}^{I} - \Sigma_{\alpha})E_{\alpha} + \sum_{\alpha' \neq \alpha} G_{\alpha\alpha'}^{I} E_{\alpha'} = 0, \qquad (2.16)$$

where the definition of all the terms remains the same as before
and the unknown amplitudes are given by $E_{\alpha} = 2iB_{\alpha} \sin(q_{\alpha}h)$ (note
that, as expected, $E_{\alpha}' = 0$). This result agrees with our interpretation
of Equations 2.12. Now, only the interface between regions I and II
is relevant, which removes the equations related to the II–III interface and the coupling between interfaces through the perforations
(which are blind), yielding $G_{\alpha}^{V} = 0$.

Although the equations presented up to here are general and can
be applied to both 1D and 2D geometries, it is worth commenting
briefly how our approach is simplified when treating the 1D case
with $k_y = 0$. In this case, it can be shown that light polarizations are
decoupled, which permits the independent treatment of *s*- and
p-polarized waves. Thus, when studying spoof SPPs in 1D indentations, we consider only *p*-polarization, as the appearance of nonzero solutions in Equations 2.12 and 2.16 is linked to the divergent
behavior of $G_{\alpha\alpha'}$ when a *p*-polarized plane wave goes grazing.
Therefore, we can restrict the expansion basis in our theoretical
framework to only *p*-polarized modes when treating 1D systems,
which considerably simplifies the calculations.

Spoof SPP modes supported by perforated metals are given by
the nonzero solutions of Equations 2.12 and 2.16. Specifically, the
dispersion relation of these bound modes can be calculated by
finding the parallel wave vector, $k_{\parallel}$, and frequency, f, which vanish
the determinant associated to the matching equations. This pro-
blem must be solved numerically in general but analytical expressions can be obtained by using two approximations:

- Only the fundamental waveguide mode (that we denote as $\alpha = 0$)
  is taken into account in the modal expansion inside the indentations, minimizing the size of the set of matching equations.

- Only the zero-order diffracted (*p*-polarized) plane wave is considered in $G_{\alpha\alpha'}$, which provides us with simple expressions for this term.

The first approximation leads to accurate results for subwavelength indentations, failing when the cross section of the indentations is comparable to the size of the array unit cell. The second approximation, which is equivalent to consider the perforated structure as an homogenous metamaterial penetrated by the average EM fields, must be corrected close to the band edges, as it does not reflect the presence of band gaps due to diffraction effects.

Let us first consider the analytical results obtained from those two approximations for the case of periodic blind perforations. The condition for the existence of bound modes in these systems is given by

$$G - \Sigma = 0, \tag{2.17}$$

where we have made $G \equiv G_{00}^{I}$ (Equation 2.15) and $\Sigma \equiv \Sigma_0$ (Equation 2.13). Neglecting diffraction effects, the G-term reads $G = iY_{k_{\parallel},p}|S|^2$, where $S = \langle k_{\parallel},p|0\rangle$ is the overlap between the p-polarized zero-order diffracted mode and the lowest indentation waveguide mode. For apertures much smaller than the wavelength, the dependence of S on the parallel wave vector can be neglected. Imposing $k_{\parallel} = k_{\parallel}u_x$, we can write the dispersion relation of the spoof surface plasmon modes as

$$k_{\parallel} = k_0\sqrt{1 + \frac{|S|^4}{\Sigma^2}}. \tag{2.18}$$

Note that, as expected, $k_{\parallel} > k_0$ that reflects the confined nature of the modes.

We now focus on metal slabs fully pierced by periodic indentations. The symmetric character of these structures with respect to the $z = h/2$ plane enables us to rewrite Equations 2.12 into a single approximate equation of the form

$$(G - \Sigma) \pm G^V = 0, \qquad (2.19)$$

where $G^V \equiv G_0^V$. Importantly, the presence of the negative (positive) sign in Equation 2.19 indicates the symmetric (antisymmetric) character of the modes with respect to the slab center. Following the same procedure as before, the analytical spoof surface plasmon bands can be now written as

$$k_\parallel = k_0 \sqrt{1 + \frac{|S|^4}{(\Sigma \pm G^V)^2}}. \qquad (2.20)$$

Again, we have $k_\parallel > k_0$ (bound modes). Equation 2.19 indicates that the frequency shift between the spoof SPP dispersion relation for partially and fully perforated metals is governed by the ratio $G^V/\Sigma = 1/\cos(qh)$, which only depends on the propagation constant of the indentation waveguide mode, q, and the thickness of the metal slab, h.

3 Spoof surface plasmons in flat geometries

In this section we study spoof SPPs in various flat configurations. We first consider in Section 3.1 flat surfaces decorated with periodic arrays of straight and slanted grooves (1D) and square dimples (2D). Next, we study in Section 3.2 slabs perforated with periodic indentations, slits (1D) and holes (2D). For the sake of simplicity, we take $\epsilon_{II} = 1$ in Sections 3.1 and 3.2. Finally, in Section 3.3 we present an effective medium approach that allows us to interpret these structures as plasmonic metamaterials.

3.1 Textured surfaces

We analyze the EM modes bound to textured PEC surfaces by considering the two simple geometries for which the spoof SPP concept was first developed: 1D arrays of grooves and 2D arrays of square dimples [14, 15]. We have shown that guided modes in these two structures are given by the nonvanishing solutions of the matching equations describing the fields continuity at the metal–vacuum interface.

First we consider the case of 1D arrays of grooves. The inset of Figure 3.1(a) shows the geometrical parameters of the structure: the array period, d, and the grooves' width and depth, a and h, respectively. As mentioned before, PEC approximation makes all lengths scalable. Thus, from now on, we take the system periodicity, d, as the unit length. The approximate spoof SPP bands obtained from Equation 2.18 reads

$$k_{\parallel} = k_0 \sqrt{1 + \left(\frac{a}{d}\right)^2 \tan^2(k_0 h)}, \tag{3.1}$$

where we have used that $S = \sqrt{a/d}$ for very narrow 1D apertures and that the propagation constant for the lowest waveguide mode inside the grooves is is $q = k_0$. This expression shows clearly the geometrical origin of the modes and allows us to predict the dependence of the spoof SPP bands on the width and depth of the indentations. Note that, according to Equation 3.1, enlarging the groove depth translates into the deviation of the dispersion relation from the light line towards larger wave vectors, and that when $a = 0$ or $h = 0$ (flat surface), $k_{\parallel} = k_0$, and no confined modes are supported by the structure.

Panel (a) of Figure 3.1 displays the normalized frequency (d/λ) versus wave vector $(k_{\parallel}d/2\pi)$ for the spoof SPP modes, in groove arrays of width $a = 0.2d$ and depths ranging from $h = 0.2d$ to $h = d$. As predicted by Equation 3.1, the bands shift to lower frequencies when the depth of the grooves is increased. This result can be understood in terms of cavity resonances occurring inside the indentations. Note that the lowest waveguide mode supported by 1D apertures has a TEM character and is always a propagating one. Thus, EM fields explore completely the groove depth, making them strongly dependent on h. This is reflected through the tangent function in Equation 3.1, which diverges when the Fabry-Perot condition, $\sin(k_0 h) = 0$, is satisfied. We can interpret this behavior as resulting from the fact that spoof SPPs in 1D blind

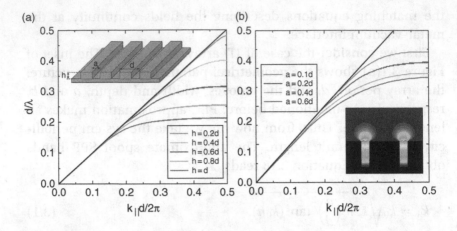

Figure 3.1 Dispersion relation of the spoof SPPs supported by periodic groove arrays. Inset of panel (a) sketches the structure considered. Panel (a) shows the dependence with h for a fixed groove width $a = 0.2d$. Panel (b) plots the bands for $h = 0.6d$ and different grooves widths. Inset of panel (b) renders the electric field amplitude for $h = 0.6d$ and $a = 0.2d$ evaluated at the band edge $(k_\parallel = \pi/d)$.
Reproduced from [101].

indentations have a hybrid nature with characteristics of both surface and cavity EM modes.

Figure 3.1(b) shows the dependence of the dispersion relation on the grooves width (a) for $h = 0.6d$. Four grooves sizes between $a = 0.1d$ and $a = 0.6d$ are considered. The mode frequency shifts to the red with larger a. This is also predicted by Equation 3.1, where $k_\parallel$ grows linearly with a far from the light line. As the ratio a/d controls the overlap between the zero-order diffracted wave and the first waveguide mode, we can conclude that the EM coupling at the interface is larger for wider indentations, which lowers the mode frequency, thus increasing the binding of the fields. The inset renders the electric field amplitude for a groove array with $a = 0.2d$ and $h = 0.6d$ evaluated at band edge $(k_\parallel = \pi/d)$. It decays more rapidly into the vacuum than inside the grooves, which agrees with the interpretation of the spoof surface plasmons as hybrid modes between surface and cavity modes.

The coupled mode scheme presented above can be extended to slanted perforations. Here, for simplicity, we will consider the case of slanted grooves [104]. This can be done by simply considering two different coordinate systems above the PEC surface (unprimed, region I) and inside the perforations (primed, region II). These are rotated with respect to each other by θ, which corresponds to the slant angle for the grooves; see bottom-right inset of Figure 3.2. The modal expansion in each region is performed in the same way as in the case of straight corrugations. The continuity equations at the slit openings now read

$$E_x^I(x,0) = \cos\theta E_{x'}^{'II}(\cos\theta x, h - \sin\theta x) +$$
$$+ \sin\theta E_{z'}^{'II}(\cos\theta x, h - \sin\theta x), \tag{3.2}$$

$$H_y^I(x,0) = H_y^{II}(\cos\theta x, h - \sin\theta x). \tag{3.3}$$

Note that they are written in terms of only unprimed spatial coordinates.

The x-dependence of Equations 3.2 and 3.3 can be removed through the standard projection procedure, yielding a set of algebraic equations of the form $\sum_{m'}(G_{mm'} - \epsilon_{mm'})E_{m'} = 0$. Note that, contrary to the straight grooves case, the matrix $\epsilon_{mm'}$ is not diagonal, and couples modes with different m once the perforations are slanted. An approximate expression can be obtained for the slanted spoof SPP dispersion relation in the form of Equation 3.1 by replacing the grooves width, a, by an effective parameter $a_{\text{eff}}(\theta) = a[1 + (\pi^2/6)(a\tan\theta/\lambda)^2]$ [104]. This result indicates that it is the actual dimension of the grooves, h, and not their depth along the direction normal to the PEC surface, $h\cos\theta$, which controls the spoof plasmon characteristics.

Figure 3.2 plots the dispersion relation of the spoof plasmon modes supported by a periodic array of grooves of width $a = 0.4d$ and height $h = 0.4d$, and three different orientations θ. By slanting the grooves, the dispersion relation shifts to higher frequencies.

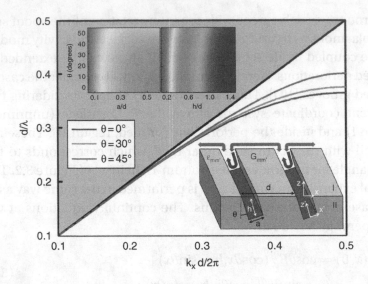

Figure 3.2 Spoof SPP dispersion relation for an array of grooves of dimensions $a = 0.4d$ and $h = 0.4d$, and three different orientation angles θ. The black line plots the light line, $k_0 = k_x$. Bottom inset: Sketch of a slanted groove array supporting spoof SPPs, together with the physical interpretation of the terms emerging in their modal expansion description. Top-left inset: Normalized spoof SPP frequency, d/λ, evaluated at the band edge ($kx = \pi/d$) as a function of the slant angle and the groove width for $h = 0.4d$. Top-right inset: The same as before but as a function of the groove height for $a = 0.4d$. In both insets the linear color scale ranges from 0.15 (blue) to 0.43 (red).
Reproduced from [104].

This trend is predicted by Equation 3.1, where the widening of a_{eff} with θ leads to a faster growth of the modal frequency with increasing parallel wave-vector. The top-left (right) inset of Figure 3.2 displays the normalized frequency, d/λ, evaluated at the spoof SPP band edge as a function of the slant angle and the grooves width (height) for $h = 0.4d$ ($a = 0.4d$). In both panels, colors render d/λ in a linear scale from 0.15 (blue) to 0.43 (red). Again, we can observe that increasing the slant angle blue-shifts the normalized frequency, and moves the spoof plasmon bands closer to the light line. The left inset of Figure 3.2 indicates that this trend is more pronounced for wide grooves $(a \gtrsim 0.4d)$ than for narrow ones, which can be also explained in terms of $a_{eff}(\theta)$. This grows as the square of $a \tan \theta / \lambda$ and hence the spoof SPPs sustained by wide indentations are more sensitive to variations in their orientation.

We now analyze the characteristics of the spoof SPP modes supported by 2D dimple arrays. For simplicity, we consider only the case of square arrays of square apertures, see inset of Figure 3.3 (a). The geometry of the system is now given by the array period, d, taken as a reference length, and the dimples side and depth, a and h, respectively. The analytical expression for the dispersion relation of the modes has the form

$$
k_{\parallel} = k_0 \sqrt{1 + \left(\frac{2\sqrt{2}a}{\pi d}\right)^4 \frac{k_0^2}{(\pi/a)^2 - k_0^2} \tanh^2\left(\sqrt{(\pi/a)^2 - k_0^2}\,h\right)},
$$

(3.4)

where we have used that $S = (2\sqrt{2}/\pi)(a/d)$ in the limit of deep subwavelength indentations [15] and that the propagation constant for the lowest waveguide mode (TE_{11}) supported by 2D square apertures is $q = i\sqrt{(\pi/a)^2 - k_0^2}$. Note that, in the range of validity of our approximate approach, $\lambda \gg 2a$, q for the TE_{11} mode is imaginary and the fields decay evanescently within the perforations. This evanescent character of the fields is reflected into the hyperbolic tangent dependence on h that Equation 3.4 yields for the spoof SPP dispersion relation. We can anticipate that the distinct character (propagating/evanescent) of the fundamental waveguide modes supported by 1D and 2D indentations leads to fundamental differences between the bound modes supported by these two geometries.

Panel (a) of Figure 3.3 plots the exact spoof SPP bands for dimples of side $a = 0.6d$, as calculated by seeking for zeroes of the determinant of the whole system of equations 24. The depth of

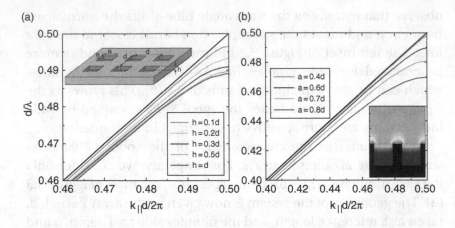

Figure 3.3 Dispersion relation of the spoof SPPs sustained by periodic dimple arrays. Inset of panel (a) shows a schematic picture of the system. Panel (a) plots the bands for $a = 0.2d$ and several dimple depths h. Panel (b) renders the bands for $h = 0.6d$ and different a. Inset of panel (b) displays the electric field amplitude at the band edge for the case $h = 0.6d$ and $a = 0.6d$. Reproduced from [101].

the indentations is varied from $h = 0.1d$ to $h = d$. As in Figure 3.1(a), the mode frequency is lowered when the depth of the perforations is increased. However, the displacement of the bands is much smaller than in the 1D case. This difference is related to the evanescent nature of the fields within the dimples and can be understood through Equation 3.4, where the tanh function leads to a little sensitivity of the modes to variations in h.

In Figure 3.3(b), the dependence of the spoof SPP bands on the dimple area is analyzed. Dimples of depth $h = 0.6d$, and sides between $a = 0.4d$ and $a = 0.8d$, are considered. As predicted by Equation 3.4, the dispersion relation bends at lower frequencies when a is enlarged. Similarly, to groove arrays, this effect is due to an increase in the EM coupling of diffracted and dimple waveguide modes, which in our analytical approach is proportional to the ratio a/d. The inset renders the electric field amplitude evaluated at the band edge for $a = 0.6d$ and $h = 0.5d$. As predicted by Equation 3.4, the electric field is mostly located at the system interface ($z = 0$) and decays into both the indentations and the vacuum superstrate. The propagation and confinement of THz radiation in flat copper surfaces pierced by square arrays of square dimples have been experimentally analyzed [82], obtaining results in accordance with the analytical results described above. Dimple arrays have also been employed to achieve a localized state of spoof SPPs at a point defect in the subwavelength scale [105] as well as guiding through coupled defect modes [106].

On the other hand, groove arrays have been utilized in the design of spoof plasmon analogues of metal insulator metal waveguides [90]. These are based on the coupled spoof modes across two corrugated interfaces, and the geometrical parameters offer a large degree of control over the properties of EM modes propagating along that waveguide. Furthermore, it has been shown that changing the geometry of the grooves gives rise to strong attractive or repulsive classical optical forces generated by such modes [107]. On the other hand, the study of fluctuation induced interactions (see Reference [108] for a review) in these kind of systems [109] has shown that the near-field radiative heat transfer between two gold plates can be greatly enhanced owing to spoof surface plasmons. [110].

3.2 Perforated slabs

Let us now study the modal properties of the spoof SPPs supported by periodic arrays of 1D and 2D apertures drilled in PEC slabs of thickness h and compare them with those of textured surfaces. As in the case of blind indentations, it is possible to construct analytical expressions for the dispersion relation of the modes by recalling Equation 2.20. Thus, for 1D arrays of slits of width a, we have

$$k_{\parallel}^{\pm} = k_0 \sqrt{1 + \left(\frac{a}{d}\right)^2 \frac{\sin^2(k_0 h)}{[\cos(k_0 h) \pm 1]^2}}, \tag{3.5}$$

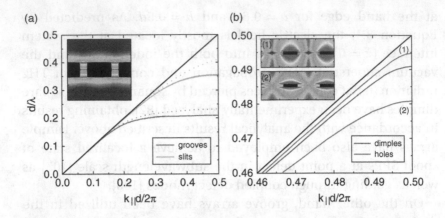

Figure 3.4 Dispersion relation of the spoof SPPs supported by fully perforated films. Panel (a): comparison between 1D slits and grooves of the same size ($a = 0.2d$ and $h = d$). Dotted lines show the analytical band obtained from Equations 3.1 and 3.5. Inset depicts the electric field amplitude at the band edge for the slit array. Panel (b): comparison between 2D arrays of holes and dimples with $a = 0.6d$ and $h = 0.3d$. Insets render the fields the edges of the two spoof SPP bands for the hole array. Reproduced from [101].

where the negative (positive) sign corresponds to bound modes whose parallel component of the electric field is symmetric (anti-symmetric) with respect to the middle plane of the film. Note that the overlap and the propagation constant remain the same as in Equation 3.1.

Panel (a) of Figure 3.4 renders the dispersion relation of the spoof SPP modes supported by slits and grooves of the same dimensions. The width and depth of the indentations are $a = 0.2d$ and $h = d$, respectively. For this set of geometrical parameters, both structures sustain only one bound mode (antisymmetric), which is tightly confined to the metal slab. Note that for thicker films, the second higher energy mode (symmetric) also lies below the light line (not shown here). In Figure 3.4(a), the dispersion relation for the slit array is raised with respect to that for the grooves, which indicates that the modes are less bounded to the structure. This is a direct consequence of the propagating character of the waveguide modes in 1D apertures, which leads to the bouncing of the fields inside the perforations. Whereas the bottom of the grooves acts as a mirror, slit openings allow the coupling to diffracted waves. This fact blue-shifts the spectral position of the cavity resonances and permit spoof SPP modes to extend out of the structure. Dotted lines plot the analytical bands for both systems. The inset of the panel displays the electric field amplitude at the band edge for the fully perforated slab. Note that the electric field vanishes at the center of the slits, showing the odd parity of the mode with respect to the middle plane of the film. Finally, we note that double slit arrays (not discussed here) have been suggested as a platform to mimic electromagnetically induced transparency [111].

Similar results to those previously presented have been obtained by different groups [112–115]. Making use of the similarity between these confined modes and the waveguide modes supported by a dielectric slab, these works have been focused on obtaining the effective dielectric response of the 1D periodic metallic structure. It has been demonstrated that, in the metamaterial limit (wavelength much larger than the period of the array), a 1D array of slits behaves as an effective dielectric medium with anisotropic parameters [113]. The effective EM parameters will be detailed in Section 3.3.

By introducing S and q for 2D apertures in Equation 2.20, the dispersion relation of the spoof SPP modes supported by 2D holes fully piercing a metal slab can be calculated. For the simple case of square holes it reads

$$k_{\parallel}^{\pm} = k_0 \sqrt{1 + \left(\frac{2\sqrt{2}a}{\pi d}\right)^4 \frac{k_0^2}{(\pi/a)^2 - k_0^2} \frac{\sinh^2\left(\sqrt{(\pi/a)^2 - k_0^2}h\right)}{\left(\cosh\left(\sqrt{(\pi/a)^2 - k_0^2}h\right) \pm 1\right)^2}}.$$

$$(3.6)$$

Note that, similar to dimple arrays, the evanescent character of the EM fields inside the apertures is reflected in the appearance of hyperbolic functions describing the dependence of the mode properties on h.

Similar to blind indentations, the distinct behavior of EM fields within 1D and 2D apertures trespassing the metal structure gives rise to different mode properties for these two systems. Whereas in 1D geometries the full perforation of the metal slab blue-shifts the mode frequency, in 2D perforations, this effect leads to the splitting of the spoof SPP band into two. This is clearly shown in Figure 3.4 (b), which plots the dispersion relation for dimples and holes of the same dimensions ($a = 0.6d$ and $h = 0.3d$). Note that, whereas the former supports only one bound mode, two different modes are sustained by the latter.

The origin of the two spoof SPP modes supported by hole arrays is clarified in the insets of Figure 3.4(b). They depict the electric field amplitude at $k_{\parallel} = \pi/d$ for the two spoof SPP bands. In both cases, EM fields are strongly localized at the film surfaces and decay into the holes. The two modes emerge from the interaction through the holes of the evanescent tails of the surface EM modes at each side of the film. The field patterns show that the lower (higher) band corresponds to bound modes having an even (odd) parity with respect to the symmetry plane of the perforated slab. Note that this phenomenology is similar to that observed in long and short range SPPs in thin metallic films [116].

It is worth analyzing the dispersion relation of these spoof SPP modes in the limit $h \to \infty$ (i.e., when the perforated PEC film is thick enough). In this limit, the coupling between the two sides of the PEC film vanishes and the two modes merge into one, whose dispersion relation reads

$$k_{\parallel} = k_0 \sqrt{1 + \left(\frac{2\sqrt{2}a}{\pi d}\right)^4 \frac{k_0^2}{(\pi/a)^2 - k_0^2}}. \tag{3.7}$$

All the calculations presented in this section have been done within the PEC approximation and, therefore, the propagation length of the corresponding spoof SPP modes is infinite. This length is reduced to finite values in a real metal due to absorption. For instance, Shen and coworkers [117] have studied the propagation length of geometrically modified SPP modes supported by arrays of grooves in the THz regime. They reported that the loss damping experienced by the spoof SPP modes is higher at larger frequencies.

3.3 Spoof Surface Plasmon Metamaterial Parameters

A further step beyond the models above can be taken by means of an effective medium approach [14, 15]. Owing to the fact that all the geometrical dimensions considered are subwavelength, $a < d \ll \lambda_0$, a perforated PEC surface behaves as an effective medium for EM fields. That is, the EM response of a single textured PEC surface can be mapped to that of a metamaterial with homogeneous but anisotropic properties. The effective parameters for the system can be obtained by:

- Comparing the dispersion relations given by Equations 3.1 and 3.7 with those of a dielectric layer on top of a PEC surface and the canonical SPPs propagating on a *real* metal surface, respectively.
- Forcing the EM fields to propagate/decay in the same way inside the corrugated metal and in the corresponding metamaterial.

This procedure allows mapping the array of grooves considered in Figure 3.1 into an effective medium layer on top of a perfectly conducting surface, as depicted in panel (b). The effective permittivity and permeability tensors of the metamaterial layer are given by

$$\epsilon_x = \frac{a}{d}, \; \epsilon_y = \epsilon_z = \infty, \tag{3.8}$$

$$\mu_x = 1, \; \mu_y = \mu_z = \frac{d}{a}, \tag{3.9}$$

where a can be replaced for $a_{\mathrm{eff}}(\theta)$ for slanted grooves. It is clear from this effective medium picture that the dispersion relation of spoof SPPs can be tailored by means of the geometrical parameters of the surface, which is a major advantage of the spoof SPPs concept. Note that if the grooves are filled with a medium with dielectric constant $\epsilon_{II} = \epsilon_D \neq 1$, the x-component of the effective permittivity acquires the form $\epsilon_x = \frac{\epsilon_D a}{d}$. This increases the binding of the surface EM modes.

The analogy between conventional and spoof SPPs can be pushed forward by considering a surface pierced by a 2D array of holes [14, 15]. As we have seen, in the case of 2D arrays of holes the first-order cavity mode is evanescent for $\lambda > 2a$, which imposes a cutoff frequency given by $\omega = \pi c/a$. In this case the effective medium parameters of the hole array read as

$$\epsilon_x = \epsilon_y = \frac{\pi^2 d^2}{8a^2} \left(1 - \frac{\pi^2 c^2}{a^2 \omega^2} \right), \tag{3.10}$$

$$\mu_x = \mu_y = \frac{8a^2}{\pi^2 d^2}, \tag{3.11}$$

$$\epsilon_z = \mu_z = \infty. \tag{3.12}$$

Interestingly, the electric permittivity $\epsilon_{x,y}$ depends on the angular frequency, and moreover, takes the canonical Drude form with a plasma frequency equal to $\frac{\pi c}{a}$. If the holes are filled with a medium with dielectric constant $\epsilon_D \neq 1$, the x and y-components of the effective permittivity above are modified, having

$$\epsilon_{x,y} = \frac{\pi^2 d^2 \epsilon_D}{8a^2} \left(1 - \frac{\omega_p^2(\epsilon_D)}{\omega^2} \right), \tag{3.13}$$

where

$$\omega_p(\epsilon_D) = \frac{\pi c}{\sqrt{\epsilon_D} a}, \tag{3.14}$$

is the effective plasma frequency and corresponds to the cutoff frequency of the holes.

This shows that the optical response of metals can be mimicked at any frequency regime by perforating a highly conducting film.

Thereby, structured surfaces with tunable geometrical parameters can spoof SPPs, which opens the way to the transferring of all the exciting prospects of SPPs to lower frequency regimes. However, this analogy must be interpreted appropriately. Importantly, the spoof plasmon metamaterial above is anisotropic, so care must be taken about the different components of the effective dielectric constant tensor. The anisotropy is also responsible for the fact that the flat region of the dispersion curve for the spoof SPP mode appears at $\epsilon_{eff} = 0$, whereas the dispersion relation for canonical SPPs bounded to the interface between two *isotropic* media flattens at $\epsilon = -1$. The important point of the Drude formula describing the dielectric response for a perforated PEC film is that the cutoff frequency of the hole waveguide marks the separation between positive and negative values for the effective permittivity.

The effective medium description of holey metal films is based on the simple spoof plasmon dispersion relation given by Equation 3.7. Importantly, this was obtained within the two approximations described in Section 2. Namely, only the fundamental mode inside the holes is introduced in the modal expansion and only the zero-order p-polarized diffracted mode is considered. By including more modes in both regions (vacuum and inside the holes), different authors have demonstrated that the *exact* dispersion relation moves closer to the light line and strong confinement only occurs for frequencies much closer to ω_p than what the effective parameter expressions given by Equations 3.12 predicts [84, 85, 100, 118].

3.4 Experimental Realizations of Flat Spoof Surface Plasmon Metamaterials

Experimental evidence of band-bending associated with the cutoff frequency of the holes carved on metallic slabs was first reported at the microwave regime of the EM spectrum [80]. The experiments involved the analysis of the angular dependence of the transmission peaks in holey metal films infiltrated with wax. In a further development by the same group [119], the excitation of spoof SPP modes in 2D arrays of holes on perforated metals was demonstrated, also at microwave frequencies, using the classical method of prism-coupling. Simultaneously, the presence of bound spoof SPP modes decorating structured metal surfaces was verified experimentally at THz frequencies [82]. In the following, we describe briefly this experimental work, which made possible the conveying of spoof SPP abilities to the THz regime, a greatly under-explored spectral range with potentialities in many different technological areas [120, 121].

The THz spoof plasmon metamaterial samples consisted in flat copper surfaces perforated by square arrays of square dimples (see top insets of Figure 3.5. The array period was set to ~ 100 μm, and the dimples side and depth were ~ 60 μm. The coupling of p-polarized free-space THz radiation to spoof SPPs was achieved

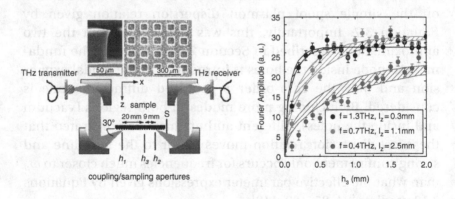

Figure 3.5 Left: Sketch of the experimental setup for the in-coupling, propagation and out-coupling of bound THz spoof SPPs in flat surfaces. The top inset shows electron micrographs of the plasmonic metamaterial sample. Right: Experimental (circles) and theoretical (shaded bands) far-field electric field amplitudes as a function of the out-coupling blade gap height for three different operating frequencies. Theoretical results were evaluated from Equation 3.15.
Reproduced from [82].

using an aperture of wavelength-scale width, defined by a blade perpendicular to the sample surface. Moreover, an intermediate aperture, h_3, was used to probe the spoof SPP decay along the direction normal to the sample surface (z-direction). Importantly, this intermediate blade also allowed distinguishing confined spoof SPP modes from weakly bound Zenneck waves [72]. A final blade was used to transform the spoof SPPs back into free-space radiation for detection.

Assuming that the intermediate blade simply blocks part of the incident spoof SPP without significantly altering its field distribution, the effect of the aperture is (in a first approximation) equivalent to integrating the Poynting vector parallel to the metamaterial surface. By expressing the spoof SPP electric field as $E(z,\omega) = A(\omega)e^{-k_z z}$ and applying EM energy conservation, the ratio between the electric field amplitudes at the input and output sides of h_3 ($A_1(\omega)$ and $A_2(\omega)$, respectively) can we written as

$$A_2(\omega)/A_1(\omega) = \sqrt{1 - e^{-2k_z h_3}}. \qquad (3.15)$$

Using that the out-coupled THz amplitude is proportional to $A_2(\omega)$, the spoof plasmon normal wave-vector, k_z, can be determined empirically from the variation of the far-field signal with h_3. The validity of the assumptions behind Equation 3.15 were tested through numerical simulations.

The right panel of Figure 3.5 plots measured THz amplitude versus h_3. Theoretical predictions based on Equation 3.15 (with k_z obtained from exact numerical calculations) are also shown. The theoretical values for the spoof SPP decay length $l_z = 1/k_z$ are indicated. The agreement between theory and the measured amplitudes is remarkable, particularly for higher frequencies where the spoof SPP confinement is strongest. This theoretical and experimental study proved the subwavelength-scale energy confinement featured by spoof SPP modes over a wide THz window.

The THz Drude-like response of perforated films has been also tested in many other configurations [83, 122, 123]. Other hole shapes different from square have also been analyzed. For example, 1D arrays of rectangular holes were shown to support the propagation of spoof SPP modes exhibiting a very low group velocity while confined in a deep subwavelength region [124, 125]. Surface EM modes supported by more complex metamaterials, such as the so-called "Sievenpiper mushrooms" were also demonstrated experimentally [126]. Other geometries, such as complementary split-ring resonators [127] or annular holes [128], were used to enlarge the operative bandwidth of the spoof SPP modes.

It is evident that one of the most appealing applications of the spoof SPP concept is waveguiding of EM signals. In addition to the examples discussed above, the molding of the flow of low frequency signals in 2D surfaces has been reported in different platforms. Thus, stopping and bending of EM radiation was achieved in spoof plasmon metamaterials by gradually varying the perforations depth [129] or width [130], or by filling them by a graded-index distribution [131]. Using similar ideas, active spoof SPP THz switches [132] were realized by placing an electro-optical material on top of the metamaterial surface. This way, active control of THz waves was demonstrated through a low-voltage electric signal.

4 Spoof Surface Plasmon Waveguides

The lack of lateral confinement of spoof SPPs in 1D and 2D indentations prevents their use as waveguides, aiming to transport EM energy within small transverse cross sections. In this section, we review a variety of waveguiding schemes based on spoof surface plasmons. We first consider spoof surface plasmon waveguides in cylindrical geometries (corrugated wires and helically grooved cylinders), including their experimental realization. Next, we discuss different spoof surface plasmon waveguides of planar character: corrugated channel, corrugated wedge, domino and conformal surface plasmon waveguides.

4.1 Spoof Surface Plasmon Waveguides in Cylindrical Geometries

Since 2004 there has been a resurgence of interest into SPP propagation along metal wires in the THz regime of the spectrum [133, 134], mostly in the context of biochemical sensing. However, the delocalized nature of the Sommerfeld waves sets constraints upon the achievable sensitivity, and leads to significant radiation loss at bends and surface imperfections. As in the case of planar interfaces, the field confinement decreases with increasing conductivity of the conductor and, in the PEC limit, metallic wires do not sustain electromagnetic surface waves anymore. The idea of increasing the binding of Sommerfeld waves through the wire corrugation was already explored in the 1950s in the context of telecommunications technology [76, 135]. These early works demonstrated that the guiding capabilities of corrugated transmission lines could be enhanced by tailoring their surface geometry. We start this section by reformulating the analysis of those two seminal works using the coupled mode method.

In this section, we apply the theoretical formalism explained in Section 2 to the spoof SPPs supported by corrugated PEC wires [86, 137], showing how the theoretical approach can be modified in order to treat these systems. Figure 4.1 shows the simplest structure supporting cylindrical spoof SPPs: a PEC wire milled with a periodic array of rings. The geometrical parameters of the system are: the wire radius, R, the array period, d, and the rings width and depth, a and h, respectively. We restrict our analysis to azimuthally (θ) independent modes, which present the lowest frequency and for which light polarizations are decoupled. This allows us to consider only p-polarized modes in our expansion basis.

As in planar geometries, we consider a unit cell of length d along the wire axis (z-direction) and divide the system into three regions: the vacuum space surrounding the wire (region I, $r \geq R$), the wire thickness occupied by the perforations (region II, $R > r \geq R - h$), and the wire core (region III, $r < R - h$). In region I, the relevant

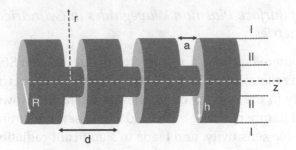

Figure 4.1 Schematic picture of the simplest structure supporting cylindrical spoof SPPs: a PEC wire of radius R perforated with an array of rings of period d, width a and depth h.
Reproduced from [136].

components of EM fields (E_z and H_θ) can be expressed in terms of p-polarized plane waves as

$$|E_z^I\rangle = \sum_n \rho_n K_0(\kappa_n r)|k_n\rangle, \tag{4.1}$$

$$|H_\theta^I\rangle = \sum_n Y_{k_n}^I \rho_n K_1(\kappa_n r)|k_n\rangle, \tag{4.2}$$

where $k_n = k_z + n\frac{2\pi}{d}$ and $\kappa_n = \sqrt{k_n^2 - k_0^2}$ are the wave vector components of the plane wave $\langle z|k_n\rangle = \exp(ik_n z)$, and $Y_{k_n}^I = ik_0/\kappa_n$ its modal admittance. Note that k_z denotes the propagation wave vector of the guided modes. The radial dependence of the fields is given by the modified Bessel functions of the second kind, K_0 and K_1 [138].

In region II, fields are only nonzero inside the rings. Thus, EM fields can be expanded as a sum over propagating and counter-propagating waveguide modes in the radial direction as

$$|E_z^{II}\rangle = \sum_\alpha D_\alpha (J_0(q_\alpha r) - \gamma_\alpha N_0(q_\alpha r))|\alpha\rangle, \tag{4.3}$$

$$|H_\theta^{II}\rangle = \sum_\alpha Y_\alpha^{II} D_\alpha (J_1(q_\alpha r) - \gamma_\alpha N_1(q_\alpha r))|\alpha\rangle, \tag{4.4}$$

where $|\alpha\rangle$ are the ring waveguide modes and $Y_l^{II} = -ik_0/q_\alpha$ their admittances. The radial dependence of the fields is now described by the Bessel and Neumann functions $J_{0,1}$ and $N_{0,1}$ [138]. Region III is filled with PEC material and therefore fields vanish within it. This provides us with a new condition into fields in region II, which must fulfill $\mathbf{E_t} = 0$ at the ring bottom. Thus, from Equation 4.3 we have $\gamma_\alpha = J_0[q_\alpha(R-h)]/N_0[q_\alpha(R-h)]$.

The matching of the EM fields at the I–II interface ($r = R$) is performed similarly as in the case of planar geometries. The z-component of the electric field must be continuous everywhere on the interface, whereas the θ-component of the magnetic field is continuous only at the rings openings. Projecting the continuity equations for the electric (magnetic) field over plane waves (ring waveguide modes), we remove the dependence on z of the matching equations. Defining the quantities

$$E_\alpha = D_\alpha(J_0(q_\alpha R) - \gamma_\alpha N_0(q_\alpha R)), \tag{4.5}$$

which correspond to the modal amplitudes of the z-component of the electric field at the rings openings, we can write the matching equations for the system in the same form as Equation 2.16.

The physical interpretation of the various terms appearing in the matching equations are the same as for planar geometries, although their expression as a function of the modal expansion coefficients is different. Thus, the $\sum$-term is now given by

$$\Sigma_\alpha = Y_\alpha^{II} \frac{J_1(q_\alpha R) - \gamma_\alpha N_1(q_\alpha R)}{J_0(q_\alpha R) - \gamma_\alpha N_0(q_\alpha R)}. \tag{4.6}$$

The G-term reads

$$G_{\alpha\alpha'} = \sum_n Y_{k_n}^I \frac{K_1(\kappa_n R)}{K_0(\kappa_n R)} \langle \alpha | k_n \rangle \langle k_n | \alpha' \rangle, \tag{4.7}$$

where the overlap integrals are defined as

$$\langle k_n | \alpha \rangle = \int dz \langle z | k_n \rangle^* \langle z | \alpha \rangle. \tag{4.8}$$

Following the same notation as for planar geometries, and as already mentioned, $\langle z|k_n \rangle$ and $\langle z|\alpha \rangle$ denote the real-space wavefunctions for the plane waves and the ring waveguide modes, respectively.

Once we have constructed the set of homogeneous matching equations, the spoof SPP modes correspond to the nonzero solutions of the modal amplitudes E_α. In Section 2 we saw that, for subwavelength perforations ($\lambda \gg a$), we can keep only the lowest-order waveguide mode in the fields expansion (note that this mode is always propagating, irrespective of the ratio a/λ). This allows us to reduce the set of matching equations to a single one, $(G_{00} - \Sigma_0) = 0$, that by introducing the expressions of the G and Σ terms reads

$$\sum_n \frac{k_0}{\kappa_n} \frac{K_1(\kappa_n R)}{K_0(\kappa_n R)} |\langle k_n|0 \rangle|^2 = -\frac{J_1(k_0 R) - \gamma_0 N_1(k_0 R)}{J_0(k_0 R) - \gamma_0 N_0(k_0 R)}, \quad (4.9)$$

where we have used that $q_\alpha = k_0$ for the TM 0 waveguide mode. The expressions for the overlap integrals $\langle k_n|0 \rangle$ can be found elsewhere [86]. In the left panel of Figure 4.2, the azimuthally independent spoof SPP bands obtained from Equation 4.9 for three different ring arrays are plotted. We take d as the unit length, the wire radius as $R = 2d$, and the ring width $a = 0.2d$. The three ring depths considered are: $h = 1.6d$ (red solid line), $h = 0.8d$ (green dotted line) and $h = 0.4d$ (blue dashed line). The dispersion relations deviate further from the light line when h is increased, in a similar way to that observed in 1D groove arrays (see panel (a) of Figure 2.1).

At low frequencies ($\lambda \gg d,a$), and for wires much thicker and rings much shallower than the array period ($R,R - h \gg d$), we can neglect diffraction orders in the G-term and obtain an analytical expression for the dispersion relation of the guided modes. By introducing the asymptotic expansions of the different Bessel functions involved in Equation 4.9, we have

$$k_z = k_0 \sqrt{1 + \left(\frac{a}{d}\right)^2 \tan^2(k_0 h)}. \quad (4.10)$$

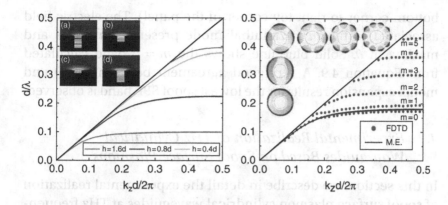

Figure 4.2 Spoof SPP dispersion relation for wires of radius $R = 2d$ perforated with rings of width $a = 0.2d$. Left panel: θ-independent bands for different ring depths. The inset shows electric field amplitude $(r > R - h)$ at the edge of the bands. Right panel: FDTD bands for higher azimuthal orders (m) for $h = 0.5d$. The insets plot the electric field patterns at the band edge ordered with increasing m from the bottom-left corner to the top-right corner of the figure. Reproduced from [136].

Note that Equation 4.10 coincides with Equation 3.1), which gives the spoof SPP bands for 1D arrays of grooves of width a and depth h. This agrees with the fact that, as in the planar case, the key parameter governing the binding of the mode is the depth of the rings, h (see Figure 4.2).

The right panel of Figure 4.2 renders the dispersion relation of the spoof SPPs on a wire of radius $R = 2d$ perforated by periodic rings with $a = 0.2d$ and $h = d$. In this calculation, there is no restriction regarding the azimuthal dependence of the EM fields. The dispersion relations (red dots) have been obtained by means of 3D finite difference time domain (FDTD) simulations. The different bands (labeled with index m) correspond to the different azimuthal symmetries of the electric field amplitude shown in the insets of the figure. For the structure considered, m ranges from $m = 0$ (θ-independent modes) to $m = 5$ (see insets from left bottom corner to right top corner of the panel). The electric field associated to the mth azimuthal mode presents $2m$ nodes and maxima in θ. Solid blue line shows the $m = 0$ band calculated from Equation 4.9. A very good agreement between FDTD and modal expansion results for the lowest spoof SPP band is observed.

4.2 Experimental Realization of THz Cylindrical Waveguides Based on Spoof Surface Plasmons

In this section, we describe in detail the experimental realization of spoof surface plasmon cylindrical waveguides at THz frequencies. We start with the milled wire structure presented in Section 4.1, and then consider a conical milled wire that concentrates EM energy at its end. Finally, we discuss the case of a helically grooved wire.

Let us first consider the spoof surface plasmon propagating along the milled wire shown in Figure 4.1. Tailoring the geometrical parameters of the wire geometry allows us to select the spectral range of operation of the structure. By choosing modulation sizes of the order of hundreds of microns, the guiding is optimized at the THz range. Figure 4.3 shows the propagation of EM fields along a 2 mm long corrugated wire of 150 μ m radius. The pitch of the corrugation is $d = 100$ μ m and the width and depth of the rings, $a = h = 50\mu$ m. The geometry of the ring array has been chosen so that the optimal frequencies for guiding are around $0.6 - 0.8$ THz. Panel (a) shows the dispersion relation, $f(k_z)$, for the guided modes supported by an infinite wire with the same geometrical parameters. Panels (b), (c) and (d) depict the electric field amplitude (evaluated at three different frequencies) for the finite wire illuminated by a radially polarized broadband terahertz pulse from the left. The field patterns have been obtained through finite-integration-technique (FIT) simulations. At the lowest frequency considered, $f = 0.4$ THz, the band lies close to the light line, which leads to a weak binding of fields to the structure. At $f = 0.6$ THz, EM radiation is guided more efficiently as the modes are strongly confined to the wire surface. At $f = 1.0$ THz, as the

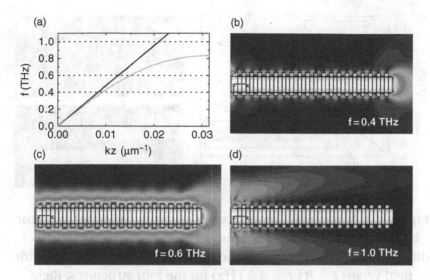

Figure 4.3 Panel (a): Dispersion relation of the guided modes traveling along a corrugated wire of radius $R = 150$ μm perforated with an array of rings of period $d = 100$ μm. The rings width and depth are 50 μm. Panels (b)-(d): Electric field amplitude for wires of length $20 \times d$ illuminated from the left by a radially polarized plane wave. Fields are evaluated at three different frequencies (0.4, 0.6 and 1.0 THz), indicated by dotted lines in panel (a). Reproduced from [136].

system does not support the propagation of any guided mode, the incident radiation is scattered out from the wire.

Taking advantage of the strong dependence of the spoof SPP confinement on the wire geometry, it is feasible to design a structure able to concentrate EM energy at one of its ends [86, 136], similar to the nanofocusing enabled by tapered plasmonic waveguides [139]. Here, we describe one candidate for this: a conical wire in which the external radius is gradually decreased along the direction of propagation, keeping the depth of the rings fixed. The left panel of Figure 4.4 plots the guided mode bands for ring arrays of period 100 μm with $h = 30$ μm and $a = 50$ μm. Four structures with wire radii ranging between 140 μm and 40 μm are

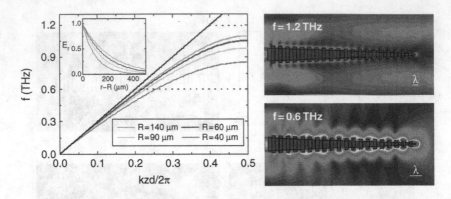

Figure 4.4 Left panel: Frequency versus propagation wave vector, k_z, for the guided modes supported by four corrugated wires of different radii. The inset plots the radial component of the electric field versus $r - R$ ($f = 0.6$ THz) for the four structures. Right panels: Electric field amplitude at 0.6 and 1.2 THz for a 2 mm corrugated cone whose radius is reduced from 140 to 40 μm. Reproduced from [136].

considered. Note that as R is decreased, the spoof surface plasmon bands deviate further from the light line. The inset of the panel renders the radial component of the electric field, E_r, as a function of the distance to the wire surface, $r - R$, for the four structures evaluated at 0.6 THz. We can see that the lowering of the bands leads to a stronger confinement of the modes with decreasing R.

The right panels of Figure 4.4 shows the electric field pattern corresponding to a 2 mm long wire whose external radius is gradually reduced from 140 to 40 μm. The structures are milled with ring arrays with the same dimensions as in the left panel. At 0.6 THz, the guided modes are tightly bounded to cylindrical wires even for the smallest radius considered. However, at larger frequencies, 1.2 THz, modes are not supported by wires with $R \leq 140$ μm. This translates into that EM radiation is guided along the cone and focused at its tip at 0.6 THz. On the other hand, the absence of guided modes at 1.2 THz makes EM waves be scattered out the structure without reaching the wire end. Remarkably, the high confinement of EM fields featured by cylindrical spoof SPPs allows the concentration of THz waves into deep subwavelength volumes in conical geometries.

Let us now consider a metallic wire periodically drilled with helical grooves. Here we present some experiments that verify the propagation of guided modes in this type of geometries. This experimental study is accompanied by a theoretical analysis of the surface EM modes supported by this complex structure [91, 140].

The experimental setup consists of a 150 mm long helically grooved wire, formed by tightly wrapping a steel wire (radius 200 μm) around a 200 μm radius core. The pitch of the resulting helical groove is $\Lambda = 400$ nm. For comparison, a bare copper wire of the same outer radius and length (600 μm and 150 mm, respectively) is also studied. Measurements are performed using time-domain THz spectroscopy. In order to discriminate the bound EM modes against unguided free space radiation, the wires are bent along the arc of a circle of radius 26 cm.

Panel (a) of Figure 4.5 displays time-domain traces of the receiver current for the wires with smooth and helically grooved surfaces. It is clear that a single-cycle-like pulse, which can be associated with a Sommerfeld wave [73], propagates along the smooth wire. However, propagation on the helical wire exhibits significant dispersion together with beating due to the presence of bound modes with different frequencies. Panel (b) plots the amplitude spectra of the traces in panel (a) together with the spectrum of a second, nominally identical sample of the helical structure which shows the reproducibility of the experimental data to small variations in optical alignment. In panel (b) of Figure 4.5, the frequency at the band edge ($k_z = \pi/\Lambda$) of the three lowest guided modes supported by the structure are indicated by vertical arrows. They are obtained by means of the theoretical FDTD and FEM calculations and correspond to the peaks in the amplitude spectra at 0.305 ± 0.002 THz, 0.326 ± 0.002 THz, and 0.353 ± 0.003 THz. In the following, we analyze the spoof SPP dispersion relation behind the structure observed in the experimental spectra.

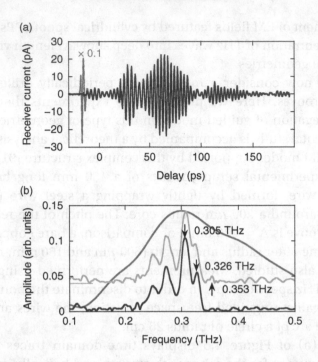

Figure 4.5 Panel (a) Receiver current as a function of time delay for the smooth wire (red line) and the grooved structure (blue line). Panel (b) Amplitude spectra of the time domain data in (a) together with the spectrum of another, nominally identical, helical sample (green curve, displaced for clarity). The arrows indicate the three azimuthal modes of the helical groove structure. The spectrum of the Sommerfeld wave (red curve) on the smooth wire extends to ∼1 THz.
Reproduced from [141].

The left panel of Figure 4.5 renders the spoof SPP bands for a helically-grooved PEC wire calculated using the FEM method [140]. The geometric parameters were chosen in accordance with experimental values. Using the notation introduced for chiral metamaterials [142], the surface EM modes can be characterized by two labels: (i) an index l, related to the azimuthal dependence, which is determined by the number of azimuthal nodes ($2l$) that the EM fields present for small values of k_z; (ii) a sign ($+$ or $-$) depending on the rotation direction at which the mode accumulates phase, $+$ for a clockwise rotation and $-$ for the counterclockwise one (as usual, the rotation direction is determined in the propagation direction). The chirality of the spoof SPPs is illustrated by the fact that for positive and negative values of k_z, the following two relations hold:

$$\omega_{\pm}^{(l)}(k_z) \neq \omega_{\pm}^{(l)}(-k_z) \tag{4.11}$$

$$\omega_{\mp}^{(l)}(k_z) = \omega_{\pm}^{(l)}(-k_z) \tag{4.12}$$

Equations 4.11 and 4.12 state that by changing the propagation direction of the EM mode, its rotation direction also has to be changed to keep the mode frequency fixed. This behavior is distinctive of helically grooved metal wires, and it is not present in either non-corrugated metal wires, nor any of the spoof plasmon platforms discussed in previous sections. It is a direct consequence of the structural chirality of the grooved wire.

The dispersion bands in Figure 4.6 also show that, in addition to the chiral degeneracy discussed above, a new type of degeneracy emerges at the band edges,

$$\omega_{\pm}^{(l)}(\pm\pi/\Lambda) \neq \omega_{\mp}^{(l+1)}(\pm\pi/\Lambda) \tag{4.13}$$

It is convenient to simplify the notation and combine the two labels into a mode index $m = \pm l$, in which the sign is related to the rotation direction and l is the azimuthal index. The evolution of the EM fields associated with the first three chiral spoof SPPs ($m = 0, \pm 1$) as their wave vector is increased from 0 to π/Λ is analyzed in the upper and intermediate right panels of Figure 4.6. To better illustrate the modal shape, we render the modulus of the real part of the E-field averaged along the propagation direction over one unit cell. Notice that the wavefront is helical within each unit cell as it follows the shape of the groove. For small values of the wave vector ($k_z = 0.3\pi/\Lambda$, upper row), the E-field patterns of the

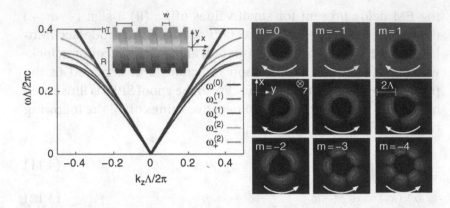

Figure 4.6 Dispersion relation of the guided modes supported by a
PEC wire of radius $R = 600$ μm inscribed with a triangular
cross-section helical groove of pitch $d = 400$ μm. The groove has
width $a = 200$ μm and depth $h = 150$ μm. The upper row of insets
displays snapshots of the electric field at the three band edges,
0.305THz (left), 0.320 THz (center) and 0.349 THz (right). The next
lower row corresponds to the first mode at 0.280 THz (left) and the
second mode at 0.180 THz (right). The pattern in the lowest row is
for the first mode at 0.180 THz.
Reproduced from [140].

chiral SPP modes with $m = -1$ and $m = +1$ are very similar.
However, close to the band edge ($k_z = 0.98\pi/\Lambda$, intermediate
row), the modes with $m = 0$ and $m = -1$ show the same azimuthal
dependence, as expected from Equation 4.13. The electric fields of
the modes $m = -2, -3, -4$ evaluated at a wave vector close to the
band edge ($k_z = 0.98\pi/\Lambda$) are displayed in the lower-right panels
of Figure 4.6. Again, the field pattern of the mode $m = -2$ closely
resembles that of the $m = -1$ mode, but with an opposite rotation
direction.

The complex k_z-dependent electric field distribution of the spoof
SPPs supported by helical grooves can be explained as follows [91,
140]. Any field component bound to a helical structure [143] can be
expanded in terms of diffracted waves as

$$F_m(r,\theta,z) = e^{ik_z z} e^{im\theta} \sum_n A_{n\ m-n}(r) e^{in(\frac{2\pi}{\Lambda}z - \theta)}, \tag{4.14}$$

where the modal amplitude $A_{n\ m-n}(r)$ contains the radial dependence of the nth-diffracted wave. $F_m(r,\theta,z)$ is an eigenfunction of the helical translation operator, $S_{\phi\,\frac{\Lambda}{2\pi}\phi}$, satisfying

$$S_{\phi\,\frac{\Lambda}{2\pi}\phi} F_m(r,\theta,z) = F_m\left(r,\theta + \phi,\ z + \frac{\Lambda}{2\pi}\phi\right) = e^{i(m+k_z\frac{\Lambda}{2\pi})\phi} F_m(r,\theta,z),$$
$$\tag{4.15}$$

where, as we have already discussed, the index m controls the symmetry properties of the EM fields. We introduce the helical coordinate $\xi = z - \frac{\Lambda\theta}{2\pi}$, which is parallel to the cylindrical coordinate z, but measured from the surface $z = \frac{\Lambda\theta}{2\pi}$. EM fields can be expressed in terms of ξ as

$$F_m(r,\theta,\xi) = f(r,\xi) e^{i(m+k_z\frac{\Lambda}{2\pi})\theta}. \tag{4.16}$$

It is now clear that this eigenfunction, evaluated along the helical surfaces ($\xi = constant$), evolves in time as $\cos[(m + k_z\frac{\Lambda}{2\pi})\theta - 2\pi ft]$, where f and t are the mode frequency and time, respectively. This simple expression explains the field profiles in the right panels of Figure 4.6, which show that the EM fields with $k_z = \pi/\Lambda$ show $2m + 1$ nodes along one helix pitch, whereas for $k_z = 0$, they show only $2m$ nodes.

4.3 Corrugated Channel Waveguides

In Sections 4.1 and 4.2 we have shown that the presence of spoof SPP modes decorating the corrugated surface of free-standing PEC wires allows for the transport of EM energy within subwavelength cross sections. However, this guiding scheme presents a major drawback: its nonplanar character makes it difficult to implement in a complex THz circuit. In this and the following sections, we present waveguide schemes based on the spoof SPP concept that feature subwavelength transverse confinement of EM fields at planar surfaces.

Here we consider corrugated channel waveguides. The design consists of corrugated V-grooves milled on a metal surface, and borrows ideas from the so-called channel plasmon polaritons (CPPs) [144] operating at visible and telecom frequencies [145]. The EM fields in un-corrugated V-grooves become more extended for increasing wavelength in such a way that, on PEC channels, they are not bound at all. Following the same strategy as in cylindrical geometries, here we review how the texturing of the metal surface leads to the emergence of bound EM modes in corrugated channels even in the PEC limit [88]. Experimental verifications of CPPs at GHz and THz frequencies were presented in References [146–148], and CPPs operating at telecom wavelengths were studied in Reference [149].

The upper inset of Figure 4.7 depicts the system under study: a V-channel of depth h and width w modulated with a periodic array of grooves of width a and depth h. The main panel shows the dispersion relation of the modes bound to the structure with $a = t = 0.5d$, $w = 0.76d$, and $h = 5d$. For this set of geometrical parameters, the angle of the channel is 20°, similar to those considered in the telecom regime. FDTD calculations demonstrate the appearance of two guided modes (from now on termed as spoof CPPs) for normalized frequencies $d/\lambda < 0.35$. Note that in corrugated V-channels of finite height, these modes present a finite cutoff frequency, as in the case of conventional CPPs [151]. Notice also the small frequency overlap between the two spoof CPP bands, which facilitates the monomode operation of the V-groove as a THz waveguide.

The longitudinal component of the electric field associated with the two spoof CPP modes is shown in the insets of Figure 4.7. The fields are evaluated at the edges of the bands. Electric fields are plotted only inside the shallow part of the corrugated V-groove. The first mode has odd parity, as the longitudinal electric fields have two lobes of different sign at both sides of the channel, vanishing at the middle plane. The second mode has even parity with respect to the symmetry plane. The tightly bounded character of the modes is clarified by introducing the modal size δ. It is

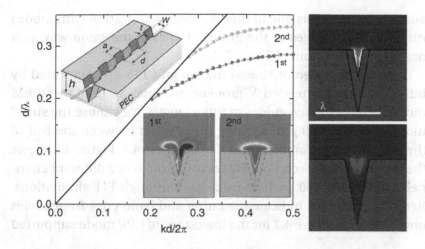

Figure 4.7 Dispersion relation of the first two spoof CPP modes supported by a corrugated V-channel milled on a PEC surface. A schematic picture of the structure is shown in the upper inset. Insets depict the amplitude of the longitudinal component of the electric field evaluated at the band edge for both modes. Right panels render the electric field amplitude at the band edge for the lowest mode evaluated at the shallower (upper panel) deeper (lower panel) sections of the channel. The horizontal white bar represents the wavelength of the mode in a vacuum. Reproduced from [150].

defined as the transverse separation between the locations where the electric field amplitude has fallen to one tenth of its maximum value, having $\delta = 0.52\lambda$, and 1.06λ for the two spoof CPPs at the band edge. The right panels of Figure 4.7 render the electric field amplitude at the band edge for the lowest spoof CPP mode. The upper (lower) panel shows the field distribution within a transverse plane located at the shallower (deeper) part of the channel, displaced $d/2$. Note that the electric field is confined into a subwavelength area, being strongly localized within the shallow section of the channel. Another interesting feature is that EM energy is not guided at the groove bottom but rather at the groove edges. We can anticipate that this is due to their strong hybridization with modes running on the edges of the groove, much in the same way as it occurs in conventional CPPs [151].

Once we have demonstrated that spoof CPPs are supported by infinitely long corrugated V-grooves, we analyze how these EM modes behave in waveguides of finite length. We choose the structure period $d = 200$ μm, keeping the relation between the rest of the geometrical parameters and d as in Figure 4.7. Figure 4.8 shows the transmission spectra of THz waves through five different channels comprising 100 periods calculated through FIT simulations. Red dashed arrows indicate the cutoff and band edge frequencies obtained from Figure 4.7 for the lowest spoof CPP mode supported

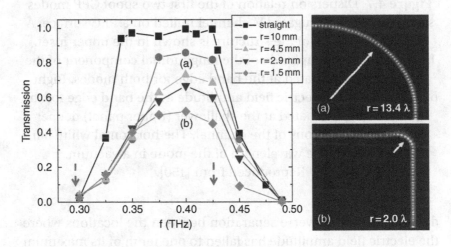

Figure 4.8 Transmission spectra for spoof CPP modes supported by corrugated PEC V-channel of period $d = 200$ μm and total length 20 mm. Straight and four different 90° bent waveguides are shown. Red dashed arrows indicate the spectral position of the cutoff and band edge frequencies in Figure 4.7. The right panels render the electric field amplitude at 0.40 THz evaluated at a height of 100 μm above the planar surface for the structures with $r = 10$ mm (a) and $r = 1.5$ mm (b).
Reproduced from [150].

by the structure. Note that, as expected, the transmission of the straight waveguide approaches unity within the spectral region (0.3–0.42 THz) between these two frequencies.

The possible use of spoof CPPs in corrugated channels for routing THz radiation requires a study of the bending losses suffered by these modes. The transmission of four 90° bends with radii of curvature r are also plotted in Figure 4.8. In these structures, $d = 200$ μm in the straight part of the channel and is slightly adjusted in the bends in order to conform with the curved geometry. For the case of maximum r (10 mm), the transmission can be as large as 90%, but is reduced as r becomes smaller, being 50% in the case $r = 1.5$ mm (around two times the wavelength). Let us stress that these bending losses are much smaller than those reported for metallic wires at THz frequencies [133]. In the right panels of Figure 4.8, the electric field amplitude at 0.40 THz for $r = 10$ mm (a) and $r = 1.5$ mm (b) in a plane located 100 μm above the planar surface is depicted. It is clear how the bending losses in these structures stem from radiation into vacuum modes occurring just at the bend of the waveguide.

4.4 Corrugated Wedges

The electric field distribution of spoof CPPs, which relates to the hybridization with modes traveling along the edges of the structure, indicates that they are not guided at the bottom of corrugated channels but at its edges (see Figure 4.7). In this section, we analyze the guided modes of a different spoof waveguiding scheme for THz waves [152].

The system under study is depicted in the inset of Figure 4.9: a PEC wedge milled with a periodic array of grooves. As in the case of corrugated channels, the EM modes supported by such geometry resemble wedge plasmon polaritons (WPPs) [153, 154] occurring at visible and telecom frequencies. The parameters defining the system supporting these EM guided modes (termed as spoof WPPs from now on) are the height, h, and angle, θ. The grooves milled on the wedge have depth t and width a, and the period of the

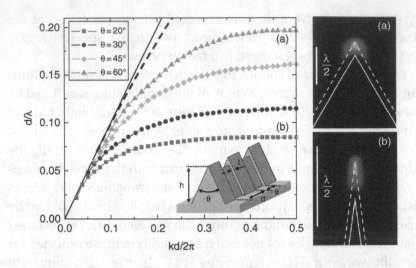

Figure 4.9 Dispersion relation of spoof WPPs traveling along corrugated wedges for different θ. Dashed black line shows the dispersion band corresponding to the flat ($\theta = 180°$) case (groove array). The inset depicts the system geometry. Right panels show the electric field amplitude at the band edge for (a) $\theta = 60°$ and (b) $\theta = 20°$. In both panels, $\lambda/2$ is represented by white bars. Reproduced from [152].

corrugation is d. In our analysis, we fix the groove dimensions, $a = t = 0.5d$, and the wedge height, $h = 5d$.

Figure 4.9 renders the dispersion relation of the fundamental spoof WPPs supported by wedges with different θ. Black dashed line plots the spoof SPP band for the limiting case of a flat ($\theta = 180°$) groove array. Reducing θ leads to the shift of the dispersion relation to lower frequencies, which implies the tighter confinement of the modes. Note that, as in the case of corrugated channels, the finite height of the structure provides the spoof WPP bands with a cutoff frequency, below which the modes lose their non-radiative nature. The right panels of Figure 4.9 render the electric field amplitude at the band edge for wedges with (a) $\theta = 60°$, and (b) $\theta = 20°$. The cross sections correspond to the deeper part of the corrugated wedge, where the EM fields are mainly localized. Remarkably, the field patterns resemble those corresponding to conventional WPPs. The half-wavelength ($\lambda/2$) is also represented by vertical white bars in both panels. The modal size for the 60° wedge is $\delta = 0.78\lambda$, whereas for $\theta = 20°$, it is equal to 0.28λ. These results demonstrate the subwavelength transverse confinement featured by spoof WPPs. Further studies on the propagation length and modal size of WPPs in the THz regime were presented in Reference [155].

As an illustrative example, we now review a functional device exploiting the propagation capabilities of these spoof WPP modes. In order to make the design work at THz frequencies, the corrugation period, d, is set to 200 μm. Figure 4.9 provides a hint on how radiation can be focused and slowed down with the aid of spoof WPPs. The lowering of the dispersion bands for decreasing θ suggests that THz waves of a given frequency propagating in a wedge which is sharpened along its length (see inset of Figure 4.10) would be gradually concentrated within the transverse plane. Additionally, THz waves at frequencies above the band edge associated with a specific θ will never reach sections of the structure sharper than that angle, being slowed down as they approach it. In order to prevent back-reflection and scattering of EM fields out of the structure, impedance mismatches along the wedge can be minimized by performing the reduction in θ adiabatically.

The inset of Figure 4.10 shows a diagram of the design proposed: a 10 mm long wedge with θ varying smoothly from 60° to 20° milled by 50 grooves disposed periodically, keeping the relation between d and the rest geometrical parameters as in Figure 4.9. The guiding properties of the structure are analyzed by means of FIT simulations under PEC approximation. Panel (a) of Figure 4.10 renders the electric field amplitude, on a line parallel to the z-axis and 100 μm above the wedge apex. Fields are evaluated at three different frequencies within the spectral range spanned by the dispersion bands shown in Figure 4.9. Waves at $d/\lambda = 0.08$ ($f = 0.12$ THz) propagate until the sharpest end of the wedge, giving rise to a maximum in the electric field amplitude located at that position (green line). At higher frequencies, radiation is slowed down and

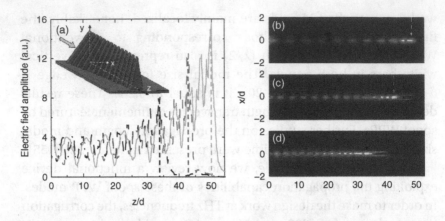

Figure 4.10 (a) Left inset: Corrugated PEC wedge with θ varying smoothly along the z-direction from 60° to 20°. Main panel: Electric field amplitude versus z along the line located 100 μm above the structure apex. Three different frequencies are considered: 0.12 THz (green line), 0.16 THz (red line) and 0.20 THz (black line). Panels (b), (c), and (d) depict the electric field amplitude within the xz plane located 100 μm above the apex for these three frequencies. Dashed arrows indicate the position of the maxima of amplitude shown in panel (a).
Reproduced from [152].

stopped before reaching the wedge end. For $d/\lambda = 0.1$ ($f = 0.16$ THz), a peak in $|E|$ is developed at the 42nd groove, for which $\theta = 26°$ (red line). At $d/\lambda = 0.13$ (0.20 THz) EM fields explore an even shorter section of the wedge and $|E|$ presents a maximum at the 34th groove, which correspond to $\theta = 32°$ (black line). Let us stress the excellent agreement of these results with the dispersion bands of Figure 4.9.

Panels (b), (c) and (d) of Figure 4.10 depict the electric field amplitude within the xy-plane located 100 μm above the wedge apex, for the three frequencies considered in panel (a). Vertical dashed arrows indicate the position of the maxima in panel (a). These three contour plots show the reduction of the effective wavelength $[\lambda_{eff} = 2\pi/Re(k)]$ experienced by the guided EM fields as they propagate along the structure. This indicates that THz waves slow down and stop at different frequency-dependent locations along the corrugated wedge. Panels (b), (c) and (d) demonstrate that guided waves are not scattered out of the wedge as they travel in the z-direction. On the contrary, while propagating along the structure, EM fields are gradually concentrated, leading to frequency selective focusing of THz waves.

4.5 Domino Surface Plasmons

Perhaps the most promising route for THz waveguiding based on the concept of spoof SPP is the domino structure (see inset of Figure 4.11(a)), as firstly introduced in Reference [89]. This structure consists of a periodic array of metallic parallelepipeds standing on top of a metallic surface, resembling a chain of domino pieces. The properties of its guided modes, the so-called *domino plasmons* (DPs), are governed by the geometric parameters defining the dominoes: periodicity (d), height of the boxes (h), lateral width (L) and inter-domino spacing a.

Here we analyze how the dispersion relation of DPs changes with the lateral width, L. To gain physical insight, as in previous cases, it is better to model the metal first as a PEC. Within the PEC approach we choose the periodicity d as the length unit. The value of a is not critical for the properties of DPs and, in these simulations, is set at $a = 0.5d$. DP bands present a typical plasmonic character, i.e., they approach the light line for low frequencies and reach a horizontal frequency limit at the edge ($k_{edge} = \pi/d$) of the first Brillouin zone (Figure 4.11(a), notice that only fundamental modes are plotted). While the limit frequency of SPPs for large k is related to the plasma frequency, the corresponding value for DPs is controlled by the geometry. In particular, the influence of the height h is clear: the band frequency rises for short dominoes ($h = 0.75d$, blue line) as compared with that of taller ones ($h = 1.5d$, black line). The most striking characteristic of DPs is their behavior when the lateral width L is changed. All bands in the range $L = 0.5d, \ldots, 3d$ lie almost on top of each other (Figure 4.11(a), grey curves). In other words, the modal effective index, $n_{eff} = k/k_0$, is rather insensitive to lateral width. Remarkably, the bands remain almost unchanged even for $L = 0.5d$, whose modal size is well inside the subwavelength regime. The described behavior is to be contrasted with that of conventional plasmonic modes in the optical regime for which sub-λ lateral confinement is not a trivial issue.

Now we study in more detail the role played by the lateral dimension L, considering now realistic metals and paying attention to the spectral regime. The periodicity d is chosen to set the operating wavelength in the desired region of the EM spectrum, and L is varied in the range $L = 0.5d, \ldots, 24d$, while the remaining parameters are kept constant ($a = 0.5d$, $h = 1.5d$). Aluminum is selected for low frequencies, where metals behave almost like PECs. In order to work at $\lambda = 1.6$ mm, providing an operating angular frequency of the order of 1 THz, we first consider $d = 200$ μm. The evolution of the modal effective index as a function of the lateral dimension normalized to the wavelength is plotted in Figure 4.11(b). The curves corresponding to a PEC (black line) and aluminum at $\lambda = 1.6$ mm (red line) are, as expected, almost identical. We can now quantify the sensitivity of the effective index to L, its variation being only about 12 % even when L goes from $L = \infty$ to $L = 0.5d = \lambda/16$, well inside the sub-λ regime. To investigate the performance of DPs at higher frequencies, the structures have been scaled down by factors $1/10$ and $1/100$. The fact that the curves corresponding to $\lambda = 0.16$ mm (green line) and $\lambda = 0.016$ mm (blue line) do not lie on top of the previous ones is a signature of the departure of aluminum from the PEC behavior. Nevertheless, even at $\lambda = 0.016$ mm, the variation of the effective index is still smaller than 15 %. When the operating frequency moves to the telecom regime ($\lambda = 1.5$ μm, magenta line) the variation of the effective index is much larger (about 38 %).

The important message of Figure 4.11(b) is that, although a variation of n_{eff} begins to be noticeable when the lateral dimension L goes below $\lambda/2$, DP bands are fairly insensitive to L in the range $L = 0.5d, \ldots, 24d$ when operating at low frequencies. Such key

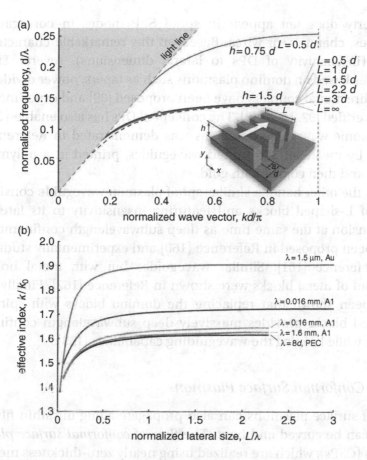

Figure 4.11 (a) Dispersion relation of DPs for various lateral widths L. Black and grey (blue) lines correspond to height $h = 1.5d$ ($h = 0.75d$). Dashed line stands for infinitely wide dominoes ($L = \infty$). Inset: diagram of the domino structure and geometric parameters (the arrow depicts the mode propagation direction). (b) DP modal effective index as a function of lateral dimension L in units of wavelength. Various operating frequency regimes are considered: $\lambda = 1.6$ mm (red), $\lambda = 0.16$ mm (green), $\lambda = 0.016$ mm (blue), and $\lambda = 1.5$ μm (magenta). To compute panel (b), a realistic description of the metals is used. As described in the main text, the periodicity d is different for the various operating frequencies, and $h = 1.5d$, $a = 0.5d$, $L = 0.5d, \ldots, 24d$.
Reproduced from [89].

property does not appear in spoof SPP modes in corrugated wedges, channels, or wires. Based on this remarkable characteristic (insensitivity of DPs to lateral dimensions), several THz devices based on domino plasmons such as tapers, power dividers and directional couplers have been proposed [89] and experimentally verified [92, 156–158]. The concept of DPs has also enabled 3D plasmonic waveguide bends, as was demonstrated in Reference [159] by means of 3D-printed waveguides, printed in a polymer resin and then coated with gold.

On the other hand, a similar spoof plasmon waveguide consisting of L-shaped blocks and featuring insensitivity to its lateral dimension at the same time as deep subwavelength confinement has been proposed in Reference [160] and experimentally studied in Reference [161]. Similar waveguides but with metal holes instead of metal blocks were shown in Reference [162]. Finally, it has been shown that replacing the domino blocks with spiral-shaped blocks provides massively deep subwavelength confinement while keeping the waveguiding capabilities [163].

4.6 Conformal Surface Plasmons

Spoof surface plasmons can also propagate along ultrathin films that can be curved and bent. The flow of *conformal surface plasmons* (CSPs), which are realized using nearly zero-thickness metal strips printed on flexible dielectric films, adapts to the curvature of the waveguide. The basic structure that supports CSPs is shown in the inset of Figure 4.12(a): a metal strip of thickness t and width W, corrugated with a periodic array of grooves of depth h, period d and groove width a. When the thickness $t \to \infty$, the structure reduces to a 1D array of grooves. Figure 4.12(a) shows the dispersion relation of the CSPs supported by this structure for different values of t. Similar to the case of the DPs, the mode dispersion is quite insensitive to the thickness of the structure. For infinite t, the dispersion relation is that of a 1D array of grooves discussed in Section 3, and the electric field points along the z-direction within the grooves, whereas the magnetic field (H-field) is directed along

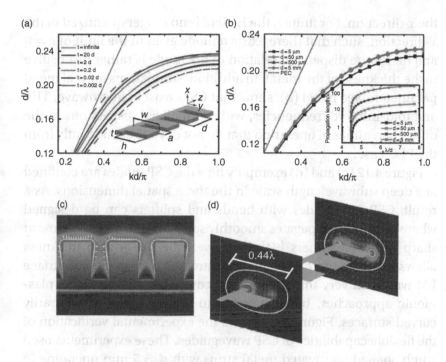

Figure 4.12 Dispersion relation and modal field patterns of the conformal plasmons. (a) Dispersion relation for the fundamental mode of the CSP waveguides for different values of their thickness, t. The geometrical parameters are $W = d$, $a = 0.4d$ and $h = 0.8d$ and the metal is modeled as PEC. The two dashed lines correspond to a double-comb structure in which the two sides of the planar metal strip are symmetrically corrugated by arrays of grooves (for $t = 0.02d$). (b) Dispersion relation of CSPs in copper at different frequency regimes with the same parameters as in panel (a) and for $t = 0.02d$. (c) Electric field amplitude in the yz plane ($z = 0$) for $d = 5$ mm and at an operating wavelength of 30 mm. (d) Power flow at two transverse planes along the CSP propagation direction for the same waveguide as in (c). Reprinted from [94].

the x-direction. For finite t, the H-field is no longer quantized in the x-direction, such that there exists a mode even in the limit of $t = 0$ and hence the dispersion relation of the mode is rather insensitive to the thickness of the metal. In addition, simulations for real metal parameters, see panel (b), show that CSPs exist at microwave, THz and mid-infrared frequencies, with normalized dispersions in the different regimes of operation that do not deviate significantly from the PEC case.

Figure 4.12 (c) and (d) exemplify how the CSP modes are confined at a deep subwavelength scale in the three spatial dimensions. As a result, CSP waveguides with bends and splitters can be designed where the mode propagates smoothly suffering little radiation loss at sharp edges or corners [94]. Moreover, their near-zero thickness allows for much more freedom to control the propagation of surface EM waves on very thin metal films compared with previous plasmonic approaches, being possible to realize CSPs on arbitrarily curved surfaces. Figure 4.13 shows the experimental verification of the flexible capabilities of CSP waveguides. These experiments used comb-shaped corrugated metal strips with $d = 5$ mm operating in the microwave regime. The waveguides were manufactured using a printed circuit board fabrication process on a three-layer flexible copper-clad laminate, consisting of a single layer of polyamide and an electrolytic copper-clad sheet connected with the epoxy adhesive. The total film thickness was 43.3 μm, such that the samples are ultra thin and flexible. As shown in Figure 4.13, they show excellent performance when wrapped around curved surfaces, and thus are well suited for incorporation into arbitrarily curved surfaces to mold the flow of CSPs.

Thus, CSPs can propagate on ultrathin and flexible films to long distances at frequencies ranging from microwave to mid-infrared. Films firstly demonstrated in Reference [94] had a thickness 600-fold smaller than the operating wavelength, and they can be bent, folded and even twisted to guide the propagation of CSPs on design. For this reason, these flexible and stretchable photonic structures can be wrapped on curved surfaces and objects, which poses advantages over devices integrated on traditional rigid

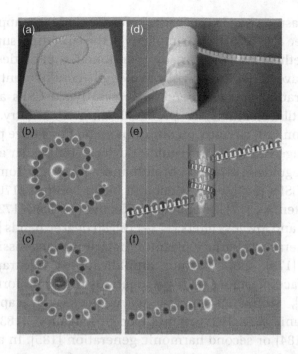

Figure 4.13 Simulation and measurement results of CSP waves on a spiral surface and a 3D helical-shaped curved surface. (a) Fabricated ultra thin sample with a spiral shape (initial and maximum curvature radius 20 mm and 40 mm, $a = 2$ mm, $d = 5$ mm, $h = 4$ mm, $w = 5$ mm, and $t = 0.018$ mm). (b) Simulation and (c) experimental electric field distributions at 11 GHz. (d) Fabricated ultrathin sample with a 3D helical shape. (e) Calculated and (f) measured electric field at 11 GHz.
Reproduced from [94].

wafer-based substrates. Furthermore, because of the deep sub wavelength confinement of the CSPs, they can propagate on curved surfaces with very low radiation losses, making them promising for applications in plasmonic circuits.

An interesting configuration is that of two coupled CSP waveguides, introduced in Reference [164]. This work showed that the strong coupling between two comb-shaped CSP structures results in spoof plasmon bands of negative refractive index for appropriate designs (see also [165]). The simplicity of the structures supporting CSPs, together with the already mentioned capabilities of this kind of waveguides (deep subwavelength confinement, broadband operation and low bending losses) place CSPs as useful and versatile elements for practical surface circuitry. Recent advances on CSP circuitry on ultrathin metal films have provided a wide range of planar waveguide devices [166], such as multiband waveguides [167, 168], broadband converters from guided waves to CSPs [169], ultra-wideband beam splitters [170], stripe planar antennas [171], frequency selective devices [172], ultra-wideband and low loss filters [173–176], 90 degree bends [177], on chip sub-terahertz surface plasmon polariton transmission lines in CMOS [178], spoof plasmon amplification [179], trapping of spoof surface plasmons with corrugations of nonuniform depth [180, 181], stopbands in CSP waveguides with capacitively coupled unit cells [182], multilayer propagation [183], beam steering [184] or second harmonic generation [185]. In addition, CSPs have also been used to suppress interference in compact spaces, which potentially solves the challenge of signal integrity [186]. Finally, CSPs propagating in structures with nonrectangular grooves have also been studied [187, 188].

5 Localized Spoof Surface Plasmons

In previous sections, we have discussed spoof SPPs that propagate both along periodically perforated planar geometries and in a variety of structured 1D waveguide configurations. Here, we discuss how textured metal particles of closed surfaces support the localized version of spoof plasmons [95, 189]. Metal cylinders corrugated with a periodic array of longitudinal grooves (see Figure 5.1) feature similar properties to the LSPs supported by metal particles in the optical regime. We start by applying the coupled mode method to localized structures in Section 5.1. Then, in Section 5.2, we show that a textured metal cylinder of size comparable to the wavelength features an EM response

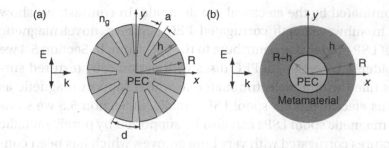

Figure 5.1 (a) A 2D corrugated PEC cylinder (invariant along the z direction) is illuminated with a wave polarized with the magnetic field along the cylinder's axis ($\mathbf{H} = H\mathbf{u}_z$). The geometrical parameters defining the structure are: radius R, periodicity d, groove depth h and groove width a. The grooves may be filled with a dielectric material of refractive index n_g. (b) In the effective medium approximation, the geometry displayed in (a) behaves as an inhomogeneous and anisotropic layer of thickness h surrounding the inner PEC cylinder.

composed of high-order modes, that is equivalent to that of a Drude cylinder in the optical regime. For this reason, such modes are termed spoof LSPs. In Section 5.3 we consider corrugated PEC particles of subwavelength dimensions, and compare its EM spectrum to that of the extensively studied subwavelength plasmonic nanoparticles.

While metal nanoparticles can be brought to the subwavelength regime by downscaling the size of the structure to the nanometer scale, the resonance frequency of the spoof LSPs modes scales with frequency. This is due to the properties of perfect metals and, as a consequence, a different strategy is needed to reach the subwavelength regime in textured PEC particles. In particular, the characteristic response of the subwavelength regime emerges by introducing a dielectric material of large refractive index within the grooves, thereby lowering the resonance frequency of the spoof LSPs modes and obtaining a dipolar response similar to the optical response of plasmonic nanoparticles at optical frequencies, which is dominated by the electrical dipole mode. In contrast, we show that in subwavelength corrugated PEC particles, novel magnetic spoof LSP modes also contribute to the spectrum. In Section 5.4 we consider subwavelength PEC disks with periodically textured surfaces filled with a dielectric material, which support magnetic as well as electric dipolar spoof LSP. Finally, in Section 5.5 we show that magnetic spoof LSPs can also be supported by purely metallic structures corrugated with very long grooves, which has been confirmed by experiments.

5.1 Coupled Mode Method for Localized Structures

Here we implement the theoretical formalism explained in Section 2 to reproduce the EM response of 2D cylinders corrugated with grooves that are periodic in the azimuthal direction. The structure, sketched in Figure 5.1(a), consists of an infinitely long metallic cylinder of radius R corrugated with a periodic array (period $d = 2\pi R/N$) of N grooves. The grooves have width a and depth h, with $(R - h)$ being the radius of the inner PEC core. The number of grooves and the radius of the cylinder are chosen to yield $d \ll \lambda_0$ (λ_0 is the wavelength of the incident wave). The grooves are filled with a dielectric material of refractive index n_g, and the surrounding medium is assumed to be air. Related structures have been studied in the past in connection with antenna design and waveguiding geometries [190, 191]. Here we make the analogy to optical plasmonics.

In order to deal with the EM response of such textured PEC cylinders, we make use of the coupled mode method. The procedure is based on the matching of the EM modes outside the cylinder (radial coordinate $r > R$, region I) and inside the grooves ($R - h < r < R$, region II) by means of the appropriate boundary conditions at $r = R$. In the following we describe the formalism technique for the structure with radial grooves, and we derive an analytical expression for the scattering cross section (SCS). In addition, we next derive the analytical spectrum of corrugated cylinders with grooves of parallel walls.

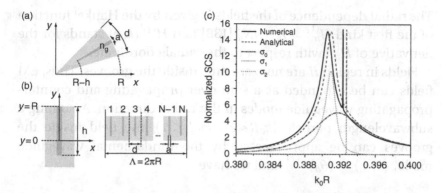

Figure 5.2 Modal Expansion technique and analytical SCS. (a, b) Unit cells used for the ME technique for the case of a cylinder corrugated with radial grooves (a) and with grooves of parallel walls (b). (c) SCS for a subwavelength PEC cylinder corrugated with radial grooves ($r = 0.4R$, $N = 40$, $a = 0.3d$ and $n_g = 8$). The solid line shows the numerically computed SCS while the red dashed line corresponds to the analytical SCS. The contributions of the $n = 0$, $n = 1$ and $n = 2$ modes to the analytical SCS are also shown

Let us first consider a PEC cylinder textured with radial grooves. In order to apply the coupled mode method to this system, we take the unit cell shown in Figure 5.2 (a). The angular size of the unit cell is $\theta_d = 2\pi/N$ while the angular groove aperture is $\theta_a = (a/d)\theta_d$. In region I, the relevant components of the EM field for p-polarized waves (**H** field pointing along the cylinder's axis) can be written as

$$|H_z^I\rangle = \sum_n \rho_n H_n^{(1)}(k_0 r)|n\rangle, \tag{5.1}$$

$$|E_\theta^I\rangle = \frac{-i}{k_0} \sum_n H_n^{(1)\prime}(k_0 r)|n\rangle, \tag{5.2}$$

with $\langle\theta|n\rangle = (\theta_d)^{-1/2}\exp(in\theta)$. In writing the above we have neglected diffraction, which is a good approximation for $d \ll \lambda_0$.

The radial dependence of the fields is given by the Hankel function of the first kind $H_n^{(1)} = J_n + iN_n$ [138] and $H^{(1)'}(k_0 r)$ stands for the derivative of $H_n^{(1)}$ with respect to the radial coordinate.

Fields in region *II* are nonzero only inside the grooves. Thus, EM fields can be expanded as a sum over propagating and counter-propagating waveguide modes in the radial direction. Assuming a subwavelength particle ($k_0 R \ll 1$, $k_0 d \ll 1$), the field inside the grooves can be approximated by the fundamental waveguide mode, $\langle \theta | \alpha_0 \rangle = (\theta_a)^{-1/2}$, and we have

$$|H_z^{II}\rangle = A\, H_0^{(1)}(k_0 n_g r) + B\, H_0^{(2)}(k_0 n_g r) |\alpha_0\rangle, \qquad (5.3)$$

$$|E_\theta^{II}\rangle = \frac{-i}{k_0 n_g^2}[A\, H_0^{(1)'}(k_0 n_g r) + B\, H_0^{(2)'}(k_0 n_g r)] |\alpha_0\rangle, \qquad (5.4)$$

where the radial dependence is in this case given by the Hankel function of the first and second kind $H_n^{(1,2)} = J_n \pm iN_n$ [138]. Finally, region III corresponds to the PEC core, where fields vanish. Hence, fields in region II must satisfy $\mathbf{E}_t = 0$ at the groove's bottom, and from Equation 5.4 we have

$$A\, H_0^{(1)'}\big(k_0 n_g[R - h]\big) + B\, H_0^{(2)'}\big(k_0 n_g[R - h]\big) = 0. \qquad (5.5)$$

Similar to the different geometries discussed in previous sections, EM fields are matched at the interface, $r = R$. While E_θ must be continuous at the interface in the whole unit cell, $-\theta_{d/2} < \theta < \theta_{d/2}$, H_z is continuous only at the groove opening, $-\theta_{a/2} < \theta < \theta_{a/2}$. Defining the modal amplitude at the grooves entrance,

$$E = A\, H_0^{(1)}(k_0 n_g R) + B\, H_0^{(2)}(k_0 n_g R), \qquad (5.6)$$

allows us to write the matching equations in the same form as Equation 2.16:

$$\left(\frac{H_n^{(1)'}(k_0 R)}{H_n^{(1)}(k_0 R)} + \frac{1}{n_g} S_n^2 \frac{f}{g}\right) E = 0. \qquad (5.7)$$

Here we have introduced the quantities

$$f = J_1(k_0 n_g R) N_1(k_0 n_g [R - h]) - J_1(k_0 n_g [R - h]) N_1(k_0 n_g R),$$

$$(5.8)$$

$$g = J_0(k_0 n_g R) N_1(k_0 n_g [R - h]) - J_1(k_0 n_g [R - h]) N_0(k_0 n_g R).$$

$$(5.9)$$

In addition, we have redefined $H_n^{(1)'}(k_0 R) \equiv \frac{dH_n^{(1)}(k_0 R)}{d(k_0 R)}$ and used the overlap integrals with the fundamental waveguide mode,

$$S_n = \langle n | \alpha_0 \rangle = \int d\theta \langle \theta | n \rangle^* \langle \theta | \alpha_0 \rangle = \sqrt{\frac{a}{d}} \operatorname{sinc}\left(\frac{na}{2R}\right). \qquad (5.10)$$

The quantity S_n contains all the information about the unit cell, and, for subwavelength texturing ($a \ll R$), its expression reduces to $S_n \approx \sqrt{a/d}$, as for the case of planar geometries.

From the matching equation system 66, the spoof plasmon modes appear as the nonzero solutions of the modal amplitude E. This yields the following transcendental equation,

$$S_n^2 \frac{H_n^{(1)}(k_0 R)}{H_n^{(1)'}(k_0 R)} \frac{f}{g} = -n_g, \qquad (5.11)$$

The solution of Equation 5.11 for a given n yields the complex resonance frequency of the mode of order n. Physically, the picture that emerges from our modeling is that of surface EM modes running around the cylinder surface, with each resonance appearing when an integer number of modal wavelengths fits into the perimeter.

In the absence of ohmic losses, the EM response of a resonant structure is characterized by means of the SCS. In order to calculate the SCS we consider (here and throughout the section) an incoming TM-polarized plane wave (magnetic field $\vec{H}$ pointing along the z direction) propagating along the x axis and with wavenumber $k_0 = \omega/c$ (see Figure 5.1). The power scattered by the structure under such illumination, P_{sc}, can be expressed as a surface integral,

$$P_{sc} = \int_{\delta V} \langle \mathbf{S} \rangle \cdot \mathbf{n} da, \tag{5.12}$$

where δV is a surface enclosing the structure. In addition, $\mathbf{S}$ stands for the Poynting vector, whose mean value reads as

$$\langle \mathbf{S} \rangle = \frac{1}{2} \mathrm{Re}\{\mathbf{E} \times \mathbf{H}^*\}. \tag{5.13}$$

The total SCS is obtained as P_{sc}/P_0, with $P_0 = c\epsilon_0 |\mathbf{E}|^2/2$ being the incident power. Normalizing to the geometrical cross section of the structure, σ_{geom}, the normalized SCS, σ, reads as

$$\sigma = \frac{P_{sc}/P_0}{\sigma_{\text{geom}}}. \tag{5.14}$$

Notice that for the 2D structures considered so far, $\sigma_{\text{geom}} = 2R$. In addition, and since we work under the PEC approximation, the resonant wavelengths scale with R, which we take as unit length. This allows us to express the frequency in terms of the dimensionless quantity $k_0 R$.

Based on the modal expansion of the fields described above, we can write an analytical expression for the normalized SCS as [95],

$$\sigma = \frac{2}{k_0 R} \sum_{n=-\infty}^{\infty} |a_n|^2, \tag{5.15}$$

where the coefficient a_n for each mode n is given by:

$$a_n = -i^n \frac{\frac{a}{d} J_n(k_0 R) f - n_g J'_n(k_0 R) g}{\frac{a}{d} H_n^{(1)}(k_0 R) f - n_g H_n^{(1)'}(k_0 R) g}. \tag{5.16}$$

In addition, we can separate the contribution of each mode as:

$$\sigma_0 = \frac{2}{k_0 R} |a_0|^2, \sigma_n = 2\frac{2}{k_0 R} |a_n|^2 \; (n \neq 0) \tag{5.17}$$

In Figure 5.2(c) we give an example of the SCS calculated with the above equations for a representative corrugated subwavelength cylinder. The analytical SCS (dashed line) shows a very good agreement with the numerically computed SCS (shown with a solid line). In addition, the analytical model allows us to present the contributions of the $n = 0$, $n = 1$ and $n = 2$ modes (σ_0, σ_1 and σ_2, respectively). As we will further discuss in the next section, our analytical model predicts that the main contributions to the SCS of 2D subwavelength cylinders come from the two lower-order modes. The magnetic and electric dipoles ($n = 0$ and $n = 1$) are very close in frequencies with the magnetic resonance being broader, and both of them contribute to the low energy peak.

The case of a PEC cylinder corrugated with grooves of parallel walls can also be treated with the coupled mode method. Given the rectangular shape of the grooves, in this case we proceed as follows in order to write the EM fields. First, we consider an array of N grooves with parallel walls on a flat surface, as shown in Figure 5.2(b). Next, we apply Born von Karman boundary conditions for a super-cell of length $\Lambda = 2\pi R$ to take account of the circular geometry. This leads to the following mode expansion in region I, under the assumption that the texturing is subwavelength

$$|H_z^I\rangle = \sum_n \rho_n H_n^{(1)}\left(k_0 y' \frac{\Lambda}{2\pi}\right)|n\rangle, \tag{5.18}$$

where $\langle x'|n\rangle = (d)^{-1/2}\exp(inx')$ and $x' = 2\pi x/\Lambda$ and $y' = 2\pi y/\Lambda$ are renormalized coordinates to account for the periodicity in the super-cell. On the other hand, the field inside the grooves reads as

$$|H_z^{II}\rangle = \left[A \exp(k_0 n_g y' \frac{\Lambda}{2\pi}) + B \exp(k_0 n_g y' \frac{\Lambda}{2\pi})\right]|\alpha_0\rangle, \tag{5.19}$$

where the fundamental waveguide mode is $\langle x'|\alpha_0\rangle = a^{-1/2}$.

By applying the boundary conditions analogously to the case of radial grooves, we arrive at the following transcendental equation for the resonance frequency of the EM mode with azimuthal number n:

$$S_n^2 \frac{H_n^{(1)}(k_0 R)}{H_n^{(1)'}(k_0 R)} \tan(k_0 n_g h) = -n_g, \tag{5.20}$$

where S_n is now given by

$$S_n = \sqrt{\frac{a}{d}} \operatorname{sinc}\left(\frac{n\pi a}{Nd}\right). \tag{5.21}$$

Note that since $\Lambda = 2\pi R = Nd$, the overlap integrals 5.21 are of the same form as for radial grooves, Equation 5.10.

Finally, and similar to the case of radial grooves, the resonance condition given by Equation 5.20 is consistent with a picture in which an EM mode runs around the cylinder surface and resonances emerge when an integer number of modal wavelengths fit into the perimeter. The tangent function in Equation 5.20 accounts for the fact that a guided mode is traveling down the groove (of depth h and refractive index n_g), and bouncing at its bottom.

5.2 Textured Metal Particles

In this section, we study the EM response of textured metal particles with empty grooves by means of the coupled mode method described in the previous section together with numerical simulations. In Figure 5.3 we characterize a representative textured PEC cylinder with radial empty grooves ($n_g = 1$) and geometrical parameters $N = 60$, $a = 0.4d$ and $h = 0.67R$. Panel (a) shows the resonance frequencies calculated with the coupled mode method detailed above in the complex plane. Each resonance frequency corresponds to the solution of Equation 5.11 for a different n. As mentioned before, this analytical model can be interpreted in a picture in which an EM mode of order n is running around the cylinder surface with resonances emerging when an integer number of modal wavelengths fit into the perimeter. The frequencies are normalized to the asymptotic frequency of the corresponding spoof SPP that would propagate on a periodically textured flat

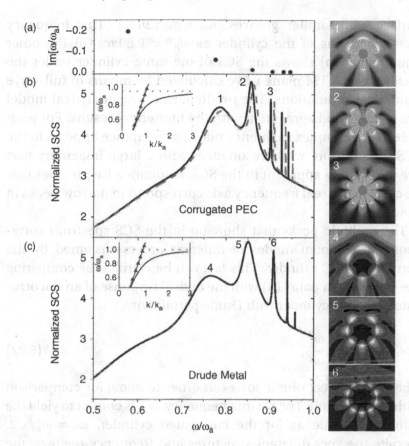

Figure 5.3 A textured PEC cylinder ($h = 0.67R$, $N = 60$, $a = 0.4d$, $n_g = 1$ and $\omega_a R = 0.89\pi c$) mimics the response a Drude metal cylinder in the optical regime. (a) Complex resonance frequencies found using the coupled mode method. (b) Numerically computed SCS (solid line) for the textured PEC cylinder. The dashed line in (b) corresponds to the calculation in the metamaterial approximation. The inset shows the dispersion relation of the corresponding spoof SPP. (c) SCS for a Drude metal cylinder of radius R with $\omega_p = \omega_a/\sqrt{2}$ and the same ω_a-value as in (b). The inset shows the dispersion relation for SPPs propagating along a Drude metal surface. The right panels display the absolute H field at the hexa-, octo- and decapole resonances for the two systems. Reproduced from [95].

surface with similar grooves, $\omega_a = \pi c/(2hn_g)$. This frequency fixes the radius of the cylinder as $\omega_a R = 0.89\pi c$. On the other hand, panel (b) shows the SCS of the same cylinder under the incidence of a TM plane wave calculated by means of full-wave numerical simulations. The predictions of the analytical model are in very good agreement with the numerical results. For each pole in the complex frequency plane, a resonance appears in the SCS. In addition, while resonances with a large imaginary part are only barely apparent in the SCS, resonance frequencies that lie close to the real frequency axis correspond to narrow peaks in the SCS.

The multiple peaks that show up in the SCS spectrum correspond to resonant modes of different orders sustained by the corrugated PEC cylinder. This fact can be clarified by comparing the spectrum in panel (b) with the optical response of an uncorrugated metallic cylinder with Drude permittivity,

$$\epsilon_m(\omega) = 1 - \frac{\omega_p^2}{\omega^2}, \tag{5.22}$$

where we neglect ohmic losses in order to allow for comparison with the PEC limit. The plasma frequency, ω_p, is chosen to yield the same $\omega_a R$ value as for the corrugated cylinder, $\omega_a = \omega_p/\sqrt{2}$. Despite the very different structures and frequency regimes, the plasmonic response of the Drude cylinder is reproduced in spectral position, value and resonance linewidth by the textured PEC cylinder. The underlying reason is the emergence of spoof LSPs that mimic the LSPs supported by metallic nanoparticles. This equivalence can be traced back to the very similar dispersion relation of the spoof SPPs and Drude metal SPPs that propagate along flat surfaces (see inset panels in (b) and (c)). Finally, the analogy between the two structures is further exemplified in the right panels of Figure 5.3, which display the field amplitude (H_z) for the first three higher-order modes that appear in the SCS (hexa-, octo- and decapole resonances) for both cases. These spoof LSP modes were experimentally demonstrated in Reference [192], while high-order radial resonances were studied and verified in Reference [193].

While the analogy between the EM response of a corrugated PEC cylinder and that of a metallic cylinder at optical frequencies is very insightful, it does not describe accurately the behavior of the EM fields inside the grooves. The reason for this is that while EM fields decay exponentially inside a real metal, the EM mode that propagates inside the grooves of a corrugated PEC cylinder leads to fields that are not evanescent. In the effective medium approximation, the corrugated cylinder can be seen as a metamaterial. As the corrugation scale is subwavelength ($d \ll \lambda_0$) the incident field does not feel the details of the structure. As a consequence, the region with grooves can be viewed as a metamaterial layer of thickness h as illustrated in Figure 5.1(b). The effective parameters can be deduced in a similar way to the case of flat spoof SPP surfaces (see Section 3). First, the PEC boundary condition in the subwavelength radial grooves implies $E_r = E_z = 0$ and $H_\theta = 0$, so $\epsilon_r = \epsilon_z = -\infty$ and $\mu_\theta = \infty$. Then, averaging $1/\epsilon$ along the θ-direction, we obtain $\epsilon_\theta = n_g^2 d/a$. Finally, as light propagates in the grooves along either the r or z directions with the velocity c/n_g, the effective material must satisfy the equations $\sqrt{\epsilon_\theta \mu_z} = \sqrt{\epsilon_\theta \mu_r} = n_g$, which yields $\mu_\theta = \mu_z = a/d$. In summary, the set of parameters in cylindrical coordinates reads as,

$$\epsilon_r = \infty, \epsilon_\phi = n_g^2 d/a, \epsilon_z = \infty, \tag{5.23}$$

$$\mu_r = a/d, \mu_\phi = 1, \mu_z = a/d. \tag{5.24}$$

In Cartesian coordinates, these effective parameters transform to permittivity and permeability tensors with off-diagonal elements and elements that depend on position. Hence, the subwavelength grooves behave as an anisotropic and inhomogeneous metamaterial layer with both electric and magnetic properties. Such metamaterial supports the propagation of spoof SPPs that present a plasmon-like dispersion relation, as shown in Figure 5.3(b) and (c).

5.3 Localized Spoof Surface Plasmons in the Subwavelength Regime

The arguably most relevant configuration in plasmonics is that of the LSPs supported by metal nanoparticles of dimensions much smaller than the wavelength of relevant light. We now study spoof localized SPs in the subwavelength regime. As we will show, corrugated PEC cylinders also feature localized spoof SPs in the subwavelength regime, similar to the dipolar LSPs that are excited in subwavelength plasmonic particles. However, and in contrast to optical plasmonics, the subwavelength regime in the spoof case cannot be reached by a downscaling of the structure, since in the PEC limit resonance frequencies scale with the size of the structure. Hence, in order to bring the spoof LSPs to the subwavelength regime, ω_a must be lowered while keeping R constant, this is, either h and/or n_g must be increased.

In this section, we consider PEC cylinders corrugated with dielectric-filled subwavelength grooves (of either radial or parallel walls). The geometry is the same as the one described in Section 5.2 (see Figure 5.1) but the grooves are filled with a dielectric material of high refractive index. This way, the resonant wavelengths, λ_{res}, can be made much larger than the size of the object and we can operate in the very deep subwavelength regime, $R \ll \lambda_{\rm res}$. By inspecting the scattering cross section (SCS) and numerical eigenmodes of such structures, we show that two different modes contribute to the lowest frequency resonance appearing in the spectra. These two modes are quasi-degenerate and, while one of them has an electrical dipole character, the second one is a magnetic dipole. We start by studying structures with radial grooves, as in Section 5.2, and then we move on to grooves of parallel walls. Both cases display very similar EM resonances, as we detail below.

Figure 5.4 presents the EM response of a textured cylinder with grooves of parallel walls filled with a material of $n_g = 8$ and with geometrical parameters $h = 0.6R$, $N = 40$ and $a/d = 0.3$. This spectrum closely resembles what is observed for very subwavelength metal particles at optical frequencies. This analogy suggests that the

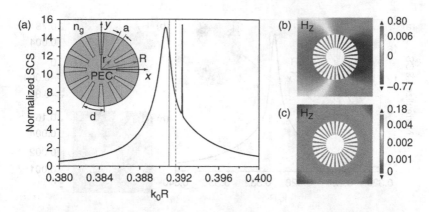

Figure 5.4 Spoof LSPs supported by a cylinder corrugated with radial grooves filled with a material of refractive index $n_g = 8$. (a) SCS spectrum. (b, c) Field pattern (H_z) at the electric and magnetic resonances. The structure has the same parameters as in Figure 5.5. The solid (dashed) vertical lines show the position of the electric (magnetic) resonance according to the analytical model.

first resonance peak is due to the electrical dipole resonance while the high frequency narrow peak is associated with the electrical quadrupole. While the assignment of the high frequency peak is correct, a calculation of the eigenmodes of the structure reveals that the first resonance peak has contributions of two different EM modes with very similar eigenfrequencies. One of them has an electrical dipole character (see panel (b)) whereas the field pattern associated with the other one is independent of the azimuthal angle, as corresponds to a magnetic dipole pointing along the z-axis (see panel (c)). In the subwavelength limit, Equation 5.11 (obtained from the coupled mode method) predicts that the mode for $n = 0$ (i.e., azimuthally-independent) is very close in frequencies to the mode for $n = 1$, but with broader linewidth. The resonance frequencies of these two modes for the textured PEC cylinder considered in Figure 5.5 are marked as vertical lines in panel (a) of that figure. As shown by these results, the analytical predictions are in a very good agreement with the numerical SCS and eigenmode calculations.

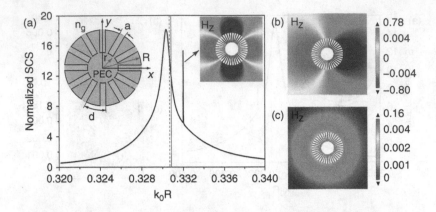

Figure 5.5 Spoof LSPs supported by a cylinder corrugated with grooves of parallel walls. (a) SCS for a structure with parameters: $r = 0.4R$, $N = 40$, $a/d = 0.3$ and grooves filled with a material of refractive index $n_g = 8$. The low-energy peak ($k_0R = 0.33$) reveals two resonances: the electrical and magnetic dipoles, whose magnetic field amplitude (H_z) are plotted in panels (c) and (d), respectively. The solid (dashed) vertical lines show the position of the electric (magnetic) resonance according to the analytical model.
Reproduced from [189].

A PEC cylinder corrugated with grooves of parallel walls displays a very similar EM response, as summarized in Figure 5.5. In that figure, we present results for a subwavelength cylinder with the same parameters as those used in Figure 5.4 but for grooves of parallel walls instead of radial. Panel (a) displays the SCS while panels (b) and (c) show the EM field patterns of the two eigenmodes that, similar to the case discussed in Figure 5.4, contribute to the first resonance peak. By comparing this figure to the case of grooves of parallel walls, it is clear that the EM response of both structures is very similar. We conclude that the only effect of the shape of the grooves is a small shift in the resonance frequencies: while for rectangular grooves the resonance is at $k_0R \sim 0.33$, for radial grooves $k_0R \sim 0.39$.

In addition, and analogously to Equation 5.11, Equation 5.20 also predicts that the modes for $n = 0$ and $n = 1$ are very close in frequencies in the subwavelength limit, as shown by the vertical lines in Figure 5.5(a). Again, we can see that, despite the few ingredients it includes, our simple modeling shows a very good agreement with the numerical SCS. Furthermore, our modeling illustrates the geometrical origin of the modes and justifies their naming: magnetic dipole LSP for the $n = 0$ mode and electrical dipole LSP for the $n = 1$ mode.

Remarkably, we can see from these results that subwavelength PEC structures textured with periodic grooves display two quasi-degenerate resonances: an electric and a magnetic dipole. This behavior is at odds to what happens in standard plasmonics. As we described in Section 1.2, the optical response of subwavelength metal nanoparticles is dominated by the electrical dipole resonance. Indeed, for these particles the magnetic dipole resonance is extremely weak and does not show up in the SCS. This fact can be understood by comparing the electric and magnetic polarizabilities provided by Mie theory in the quasistatic limit [194],

$$\alpha_e = 4\pi R^3 \frac{\epsilon(\omega) - 1}{\epsilon(\omega) + 2}, \tag{5.25}$$

$$\alpha_m = 4\pi R^3 \frac{(k_0 R)^2}{30} [\epsilon(\omega) - 1], \tag{5.26}$$

where $\alpha_{e,m}$ stand for the electric and magnetic polarizabilities, respectively, R is the radius of the particle, assumed to be in free space, and $\epsilon(\omega)$ its electric permittivity. From the above expressions, we see that $\alpha_m/\alpha_e \sim (k_0 R)^2$, and hence for plasmonic particles the magnetic response is much lower than the electric one.

5.4 Corrugated Metallo-Dielectric Disks

Here we study the 3D counterpart of the corrugated PEC cylinders studied in the previous sections. As the infinitely long 2D textured cylinder is transformed to a 3D corrugated disk of finite thickness L, the resonances shift when the particle's size is reduced. While this is the expected behavior from particle plasmonics (the dipolar resonance wavelength decreases as particle's size is reduced), the quasi-degeneracy between the electric and magnetic modes is lifted. In particular, the magnetic and electric LSP split in such a way that the former shifts to higher frequencies than the latter. In the following, we first discuss the SCS obtained for PEC disks corrugated with dielectric-filled grooves of parallel walls. Then, we compare the EM resonances supported by these structures with the electric and magnetic dipole modes displayed by subwavelength dielectric particles, focusing on the different frequency ordering of the resonances. Finally, we present a metamaterial interpretation for the corrugated PEC structures.

The SCSs for two disks of different thickness, L, are shown in Figure 5.6(a). The disks are corrugated with grooves of parallel walls filled with a dielectric material and with the same parameters considered in Figure 5.5. As in the 2D case, in the 3D FEM simulations we have considered an incident wave polarized with the magnetic field pointing in the z direction and propagating along the x axis. The SCS is calculated from the scattered Poynting vector, following Equation 5.14, where the geometrical normalization factor is in this case $\sigma_{geom} = 2RL$. Two distinct peaks can be seen in the SCS plot for disks of thickness $L = R$ (solid line) and $L = R/2$ (dashed). The field patterns at both peaks, displayed in panels (b) and (c) for $L = R/2$ case, show that the peak at a low frequency corresponds to the electric LSP while the one at a higher frequency is the magnetic LSP. This is consistent with the vectorial plots also displayed in those panels. In the first case, panel (b), the E-field arrows go from one end of the structure to the other, as corresponds to an electrical dipole, whereas in panel (c) the arrows show hoy the magnetic field circulates around the disk, as corresponds to a magnetic dipole. Hence, while infinitely long cylinders with periodic subwavelength corrugations present a magnetic LSP that is very close in frequencies to the electric LSP, finite corrugated disks support a magnetic LSP at a different frequency than the electric LSP.

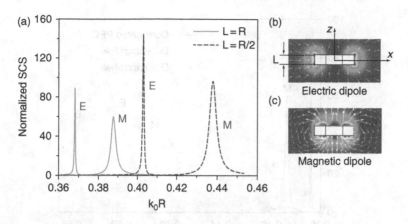

Figure 5.6 Magnetic LSPs in subwavelength PEC disks corrugated with grooves of parallel walls and filled with a dielectric. Structures of finite thickness and with the same h, N, a/d and n_g as in Figure 5.5 are considered. (a) SCS for corrugated disks of thickness $L = R$ (solid line) and $L = R/2$ (dashed line). (b, c) Near-field distribution of the electric (b) and magnetic (c) dipole resonances for a disk of thickness $L = R/2$. The arrows show the electric field lines and the color map shows the norm of the electric field in panel (b) and the magnetic field lines and magnetic field, respectively, in (c). Reproduced from [189].

Standard dielectric particles of subwavelength sizes and large refractive index also display magnetic and electric dipole modes [195, 196] reminiscent of the ones under study here. It is therefore relevant to compare the EM modes supported by corrugated metal cylinders with those emerging in subwavelength dielectric particles in the same frequency range. Figure 5.7 compares the calculated SCS for three different structures: (i) a PEC disk corrugated with grooves filled with a dielectric material of index n_g, as in Figure 5.6, (ii) a dielectric disk of radius R and refractive index given by n_g, and (iii) a dielectric shell of width h surrounding a PEC core of radius $R - h$. As opposed to the modes for the corrugated PEC disk, for dielectric particles the magnetic dipole resonance lies

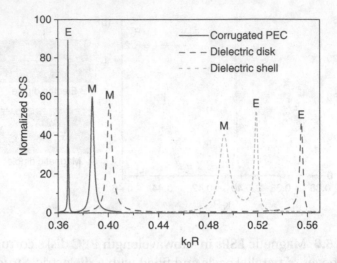

Figure 5.7 Electric and magnetic dipole resonances in subwavelength dielectric particles. The computed SCS is shown for three cases: (i) a corrugated disk of $L = R$ and with the same parameters as in Figure 5.6 (solid), (ii) a dielectric disk of the same size and $n_g = 8$ (dashed), and (iii) a PEC core of radius r with a dielectric shell of radius R and $n_g = 8$ (short dashes).

at lower energies than the electric dipole resonance, as well as for the core-shell structure. The physical origin of this opposite frequency ordering for metallic and dielectric disks is related to the dissimilar shapes of the corresponding electric field line patterns (see Reference [189] for more details). This distinct behavior implies that the corrugated PEC structure cannot be interpreted as an effective isotropic dielectric medium. Instead, in the effective medium limit, $\lambda \gg R$, the region with grooves can be mapped into a metamaterial with both anisotropic permittivity and permeability tensors, as we show below. On the other hand, it is interesting to note that, as opposed to the corrugated PEC disks, hollow textured PEC particles support a magnetic dipole at a lower frequency than the electric dipole, as studied in Reference [234].

Similar to the effective medium approximation of the 2D structure, given in Section 5.2, the 3D disks textured with dielectric-filled grooves can also be described as an equivalent metamaterial with anisotropic and inhomogeneous parameters. The effective parameters that characterize such metamaterial are the ones previously derived and given in Equations 5.23 and 5.24.

The effective medium approximation can be validated by numerical simulations. Figure 5.8 shows numerical results for the two cases depicted in panels (a) and (b): first, in the 2D case it consists of a PEC inner core surrounded by a metamaterial shell of width h (a); second, in the 3D case we deal with a shell of width h and thickness L surrounding the PEC core (b). The plot in panel (c) presents the

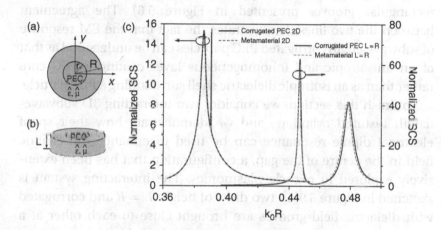

Figure 5.8 Metamaterial approximation for subwavelength corrugated PEC structures. In the effective medium approximation corrugated PEC particles behave as an inhomogeneous and anisotropic layer of thickness $R - r$ wrapped around a PEC inner core. (a) and (b) depict the geometry of the metamaterial models in 2D and 3D, respectively. (c) Numerical SCS for corrugated PEC cylinders with radial grooves ($h = 0.6R$, $N = 40$, $a = 0.8d$ and $n_g = 8$) and disks (same parameters, thickness $L = R$) together with the SCS in the effective medium approximation for both cases. Reproduced from [189].

computed SCS. In the 2D case, the plot shows the SCS for the same subwavelength corrugated cylinder considered in Figure 5.4 (solid line) together with the SCS obtained in the metamaterial approximation (dashed line). We can observe that the two lines virtually coincide. Regarding the effective medium approximation for the 3D corrugated disks of thickness L, the 3D permittivity and permeability tensors can be approximated by the 2D ones (Equations 5.23 and 5.24), which have to be implemented in a shell of width h around the inner core of thickness L. In the figure, the SCS for the corrugated PEC particle is shown as a solid line (referred to the right axis), while the effective medium approximation corresponds to the dashed line. In both cases the low energy peak is due to the electric LSP, while the high-energy peak is the magnetic LSP (similar to the results for rectangular grooves presented in Figure 5.6). The agreement between the two lines demonstrates the fact that the EM response of subwavelength corrugated PEC particles can be understood as that of an anisotropic and inhomogeneous layer coating a PEC core rather than as an isotropic dielectric shell surrounding a PEC particle.

To finish this section, we consider two interacting 3D subwavelength textured cylinders, and we demonstrate how their spoof electrical dipole resonance can be used to enhance the electric field in the centre of the gap, a configuration that has been extensively explored in particle plasmonics. The interacting system is sketched in Figure 5.9(a): two disks of height $L = R$ and corrugated with dielectric-field grooves are brought close to each other at a separation distance s. In order to maximize the field enhancement at the gap we choose a plane wave illumination propagating in the direction of the cylinders axes (z direction in the sketch) and polarized along the long axis of the structure (x direction). In this manner, the electric dipole mode of the interacting structure is excited.

Figure 5.9(b) shows the field enhancement as a function of the separation between the disks for three different values of a/d. The electric field enhancement is evaluated at the dipole resonance of the combined system and at the centre of the gap. The main result of this figure is the significant increase in field enhancement for decreasing separation (see also the inset panels), as expected from a

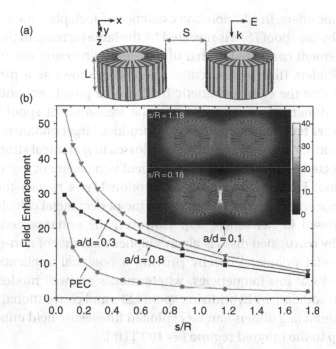

Figure 5.9 (a) Sketch of two 3D corrugated PEC disks of radius R and height L, separated by a distance s and illuminated with a plane wave polarized along the x axis and propagating along the z direction. (b) Electric field enhancement in the centre of the gap as a function of the separation for corrugated PEC cylinders with $h = 0.6R$, $N = 30$, $n_g = 8$, $L = R$ and different values of a/d. The field enhancement is evaluated at the electric dipole resonance frequency of the dimer. The dots correspond to simulation results and the lines are a guide to the eye. Circular dots correspond to a dimer of PEC cylinders without corrugation. Inset panel: electric field enhancement in the $x - y$ plane at the middle of the cylinders for $a/d = 0.1$ at two different values of s.
Figure reproduced from [95].

plasmonic effect. In addition, the existence of such plasmonic effect caused by the spoof LSPs is revealed by the large increase in the field enhancement compared to that of a system of two non-corrugated PEC cylinders (line with circular dots). This shows as a proof of concept that the electromagnetic field can be greatly enhanced in particle dimer structures thanks to the excitation of spoof LSPs, while other textured geometries may produce larger enhancements of the field. On the other hand, ohmic losses in real metal structures are expected to have an effect on the field enhancement associated with spoof LSPs. In particular, absorption losses reduce the on-resonance enhancement compared to the perfect metal calculation, as discussed in Reference [95]. However, the enhancement displayed by corrugated disks is always higher than that of non-corrugated metal cylinders, hereby proving a potential application of spoof LSPs at low frequencies, where metals are well modeled as PECs. In addition, by hybridizing spoof LSPs and conventional LSPs, these interacting dimers can be exploited for electric field enhancement up to the infrared regime (~ 100 THz).

5.5 *Magnetic Localized Spoof Surface Plasmons in Purely Metallic Structures*

In the previous sections, we showed how PEC structures corrugated with subwavelength grooves filled with a dielectric material support magnetic spoof LSPs. Although these magnetic LSP modes (and those supported by dielectric particles) are very promising in order to create dipolar magnetic resonances, they rely on dielectric materials with very high refractive indexes, which could place limitations on the feasibility of their implementation. Building up magnetic LSP modes supported by purely metallic structures would be more convenient from a practical point of view. Our analytical model detailed in Section 5.1 gives a clue as to how to proceed. Equation 5.20, which applies to cylinders corrugated with rectangular grooves, predicts that, in the subwavelength limit, the positions of the complex resonance frequencies are mainly controlled by the product $n_g \cdot h$. Therefore, an interesting design is to replace the grooves with $h \leq R$

and filled with a dielectric material by very long empty grooves that posses an effective length much larger than R. In order to achieve a depth $h > R$, the grooves of parallel walls need to be bent and warped within the particle's volume. Note that purely metallic structures displaying magnetic resonances have been explored before both in the microwave and metamaterials areas of research. Spiral structures of varied geometries have been traditionally used as frequency-independent antennas or for polarization control at low frequencies [197–199], and, more recently, spiral-based metamaterials like Swiss rolls [20] or one-spiral resonators [200, 201] have been introduced in order to improve the magnetic response of split-ring resonators [20]. Here we further exploit the concept of spoof LSPs to design metal structures with a magnetic response.

We now introduce two structures that exploit the basic idea of using very long grooves: PEC cylinders or disks corrugated with meanders and with spiral grooves. Numerical simulations confirmed that bent grooves behave like straight larger ones despite the possible internal reflections at the bends [189]. In Figure 5.10 we summarize the results on the LSP modes in PEC structures corrugated with meanders. The geometry we are considering is sketched as an inset: a cylinder of radius R is textured with four meander-shaped grooves of effective depth $h_m \approx N(R - r)/4$. The computed SCS for such a structure is plotted as a black line for the 2D case. Similar to the already discussed PEC cylinder with filled grooves, the SCS shows a dominant peak that reveals electrical dipole and magnetic dipole field patterns, as shown in panels (b) and (c). A rough estimation based on Equation 5.20 (where h is substituted by h_m) predicts spectral locations of these two dipolar modes at $k_0 R = \pi R/(2h_m) \approx 0.262$, very close to the numerical value, 0.268. Moreover, 3D structures corrugated with meanders also behave in a similar way to PEC disks corrugated with filled grooves. In particular, the SCS for disks with meanders (thicknesses $L = R$ and $L = R/2$), shows a shift in the resonant modes, with the magnetic LSP spectrally located at higher frequencies than the electrical LSP. In addition, our design allows us to create EM modes whose resonant wavelengths can be made much larger than the size of the

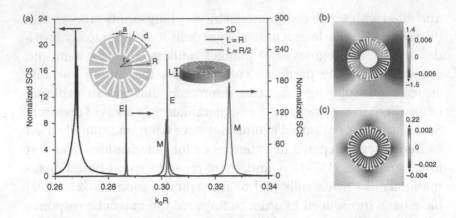

Figure 5.10 Magnetic LSPs in PEC subwavelength structures without dielectric filling. (a) The plot shows the SCS for an infinite cylinder (black line) and disks of thicknesses $L = R$ (red line) and $L = R/2$ (blue line), corrugated with four meanders and with parameters: $N = 40$, $a = 0.3d$, $n_g = 1$ and effective depth $h_m \approx 6R$. (b, c) Field pattern (H_z amplitude) for the 2D case at the dominant peak in the SCS, $k_0R = 0.268$, showing the electric dipole (b) and magnetic dipole modes (c), respectively.
Reproduced from [189].

object. This is illustrated in Figure 5.11, where the normalized SCS of an infinite cylinder corrugated with only one very long meander displays a resonance at a wavelength that is around 100 times larger than the radius. Note that similar corrugated rigid particles support Mie-like acoustic resonances that have been employed to develop deep subwavelength acoustic sensors [202].

A different way of achieving grooves of very long depth is based on a spiral geometry. As in the case of meander-shaped grooves, a PEC particle textured with spiral-shaped grooves mimics the EM response of a corrugated cylinder with shallow grooves and filled with a dielectric. The structure consists of a PEC cylinder of radius R drilled with four spiral-shaped grooves that are wrapped around a small PEC cylinder of radius r (see the geometry sketch in

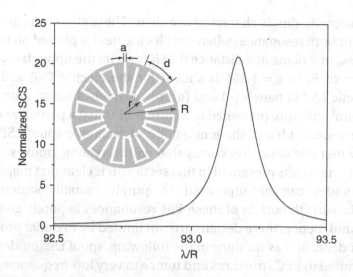

Figure 5.11 Deeply subwavelength magnetic LSP in a PEC subwavelength structure without dielectric filling. The plot shows the SCS for an infinite cylinder corrugated with just one meander and with the same parameters as considered in Figure 5.10. Reproduced from [189].

Figure 5.12(a)). Each spiral groove has width a and depth $h = 5R$, and two neighboring arms are separated by a distance d. The plot in panel (a) presents the computed SCS for 2D cylinders (left axis), and 3D disks of thicknesses $L = R$ and $L = R/2$ (right axis). As in the previously discussed cases, the dominant peak in the 2D SCS shows two quasi-degenerate electric LSPs and a magnetic LSP (H_z for these modes is depicted in panels (b) and (c)). The 3D cases present a shift of the resonances that takes the magnetic LSP mode to higher frequencies than the electric LSP. Plots of the near-field distribution at both LSP modes for a 3D disk corrugated with spiral-shaped grooves are also presented in the figure. Panels (d) and (f) show the z component of the magnetic field, H_z, at a plane cut in the middle of the disk ($z = L/2$) for the electric (d) and magnetic (f) LSPs. For both resonances, the near field resembles that of the 2D case (see panels (b) and (c)), and the electric-dipole and magnetic-dipole characters are clear. These characters are still clear for both resonances when the electric field is plotted on top of the disk, in a plane at a distance $0.25R$ from its upper face. The pattern of E_z at $z = 1.25L$ is shown for the electric LSP and the magnetic LSP in panels (e) and (g), respectively. Finally, note that the spiral structure presented in Figure 5.12 is just a particular case of the general idea of the emergence of magnetic spoof LSPs in purely metallic structures corrugated with very long grooves.

With the results presented in this section, it is clear that magnetic LSP modes can be supported by purely metallic structures. Additionally, the origin of these EM resonances is purely geometrical and, hence, these designs are not limited by material properties. Indeed, and as we show in the following, spoof LSP modes are not limited to PEC structures and hence to very low frequencies. By considering realistic metal properties, Reference [Huidobro2015] showed that these modes exist for a broad range of frequencies.

In order to illustrate the range of validity of the PEC approximation, we present calculations where the effect of both metal absorption and finite permittivity have been taken into account [96]. The metal is modeled as silver with a permittivity given by the Drude formula (parameters $\omega_p = 1.37 \times 10^{16}$ Hz, $\gamma = 2.73 \times 10^{13}$ Hz, as given in Reference [203]). The extinction cross section (ECS), sum of the scattering and absorption (ACS) cross sections, is analyzed for different frequency ranges. In particular, we focus on the 2D meanders structure presented in Figure 5.10 (a), with the PEC replaced by silver. The simulation results for particles of radius $R = 10$ mm (b), $R = 1$ mm (c), $R = 100$ μm (d), $R = 10$ μm (e) and $R = 1$ μm (f), with all the other geometrical parameters scaled accordingly, are presented in Figure 5.13. Panel (a) shows the already discussed result for PEC for comparison. The structures are resonant for ~ 1 GHz, ~ 10 GHz, ~ 100 GHz, ~ 1 THz and ~ 10 THz, respectively. Realistic metal properties have three main effects. First, due to absorption in the metal, the extremely narrow band quadrupole mode disappears from the spectrum. Second, the resonance peak is red-shifted with respect to the estimation given by the PEC approximation as the frequency is

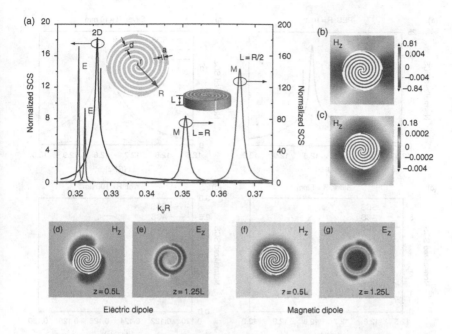

Figure 5.12 Particles corrugated with spiral-shaped grooves. (a) The plot shows the SCS for the 2D case and disks of thicknesses $L = R$ and $L = R/2$, with parameters: $d = 0.159R$, $a = 0.33d$, $r = 0.063R$, $h_m = 5R$ and $n_g = 1$. (b, c) Field pattern for the infinitely long cylinder corrugated with spirals at the two quasi-degenerate modes at $k_0R = 0.3245$ showing the electric and magnetic dipoles, respectively. (d, e) Near-field distributions of a 3D PEC disk corrugated with spiral grooves at the electric LSP resonant frequency ($k_0R = 0.32$) on two plane cuts. The magnetic field, H_z is plotted in the $x - y$ plane at $z = L/2$ (d) and the electric field, E_z, is plotted at $z = 1.25L$ (e). (f, g) The same as panels (d, e) but for the magnetic LSP ($k_0R = 0.32$): H_z at $z = L/2$ (f) and E_z at $z = 1.25L$ (g). The thickness of the disk in panels (d–g) is $L = R/2$. Reproduced from [189].

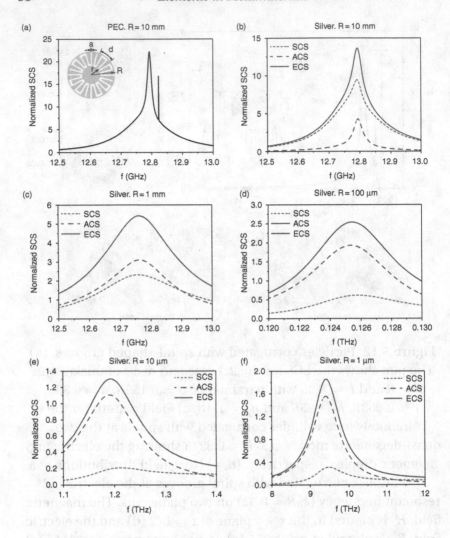

Figure 5.13 Effect of realistic permittivity in cylinders corrugated with meanders. (a) Normalized SCS for a 2D PEC cylinder corrugated with meander-like grooves (same as in Figure 5.10). The radius of the cylinder is $R = 10$ mm. (b) Normalized SCS, ACS and ECS for a corrugated cylinder made of silver and with radius $R = 10$ mm. The plot shows a very good agreement with the PEC approximation in panel (a). (c–f) Effect of the realistic conductivity for corrugated cylinders with radius $R = 1$ mm (c), $R = 100\,\mu$m (d),

raised. This shift is most noticeable for the case of $R = 1$ μm; while the real metal simulation yields a resonance at 9.5 THz, the PEC prediction is 12.8 THz. Finally, absorption starts to dominate over scattering: whereas in the GHz regime the ECS is dominated by the scattering contribution, in the IR regime absorption is the main contribution to the ECS. In conclusion, although the PEC approach loses quantitative prediction power as the frequency increases, it is able to capture the main ingredients of the resonance behavior of corrugated metal particles for frequencies ranging from the GHz to the THz and mid-IR regimes.

Next, we summarize here experimental results that confirm the existence of magnetic spoof LSPs. In Reference [189], ultrathin metal spiral structures were fabricated and characterized. For practical reasons, measurements were performed in the micro-wave regime, but, as we have discussed above, magnetic LSP modes can also be devised at higher frequencies (THz up to mid-IR energies) by just a proper down-scaling of the structure analyzed in this experimental study. The fabricated structure has an outer radius $R = 9.5$ mm and contains four metallic spiral arms wrapped 1.5 turns around a small metallic disk of radius r (a sketch of the geometry is shown in Figure 5.14). Each strip has width w and the separation distance between two neighboring arms is d. The textured disks were fabricated from copper, and with a thickness that is much smaller than the radius, $L = 0.035$ mm. This ultrathin structure is fabricated on top of a dielectric substrate using the standard printed circuit board fabrication process. The

---

Caption for Figure 5.13 (cont.)

$R = 10$ μm (e) and $R = 1$ μm (f). The geometrical parameters in all cases are $r = 0.4R$, $N = 40$, $a = 0.3d$, $n_g = 1$ and effective depth $h_m \approx 6R$. The electric permittivity of silver is modeled with a Drude formula with parameters given by

Reference [203]

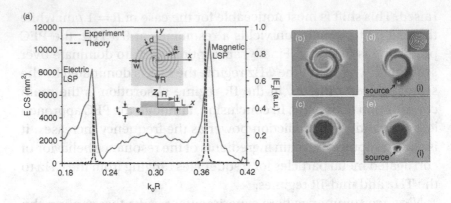

Figure 5.14 Resonance spectrum for subwavelength ultrathin textured metallic disks. (a) Experimental near-field spectrum ($|E_z|$ is measured) and calculated ECS spectrum. The inset panels show a sketch of the structure, which is placed on top a dielectric of thickness $t_s = 0.8$ mm and $\epsilon_s = 3.5$. The parameters of the fabricated structure are: $R = 9.5$ mm, $r = 0.6$ mm, $d = 1.508$ mm, $a = 1.008$ mm and $L = 0.035$ mm. (b, c) Calculated near-field distribution of the E_z field component at the electric (b) and magnetic (c) LSPs in a $x - y$ plane cut 1.5 mm above the ultrathin disk. In the color scale, red and blue indicate positive and negative values, respectively. (d, e) Measured near-field distribution of the LSP. The same quantities as in panels (b, c) are plotted. A photograph of the sample is rendered in panel (d). Reproduced from [189].

substrate has a thickness $t_s = 0.8$ mm and a dielectric permittivity $\epsilon_s = 3.5$.

The measured near-field spectrum for the ultrathin spiral structure is presented in Figure 5.14(a) with a solid line. The EM response of the structure is measured by means of a transmitting antenna that is placed 3 mm away from one side of the sample to excite the modes, and a receiving antenna that is located at the other side of the sample to detect the resonance spectrum. Two distinctive resonance peaks at frequencies $f = 1.1$ GHz

($k_0 R = 0.22$) and $f = 1.87$ GHz ($k_0 R = 0.37$) can be clearly identified. To reveal the physical nature of these two resonances, the calculated ECS spectrum for the same structure is also shown (dashed line). The calculated resonance frequencies are in excellent agreement with the experimental results. The ECS has been calculated by means of full wave simulations [204] in which both the metal absorption and the presence of a dielectric substrate are fully taken into account. The simulated field patterns at the two resonances are plotted in panels (b) and (c). Similar to the behavior previously discussed for PEC disks corrugated with meanders and spirals, these plots show that the first resonance is due to the electrical LSP (b) while the second peak emerges from the magnetic LSP (c). In addition, these field patterns closely resemble those plotted in Figure 5.12 (panels (e) and (g)) for the spiral structures of thickness $L = R/2$ considered in the previous section. The measured near-field distributions are presented in panels (d) and (e), which show the local z component of the electric field for both modes. A comparison of panels (d) and (e) to the numerical results shown in (b) and (c), shows a remarkable agreement between the simulation results and the experimental images. Hence, these experimental results show that electric and magnetic LSP resonances also exist for ultrathin ($L \ll R$) metal films and, therefore, meander and spiral structures can be used as building blocks in the design of 2D metamaterials (i.e., metasurfaces [39, 41]), as shown for instance in Reference [205].

Finally, we note here recent progress on spoof LSP particles in ultrathin platforms. Different designs have shown the versatility of particles corrugated with grooves of parallel walls, and many different geometries have been studied, including closed textured cavities [206, 207], corrugated ring structures [208, 209], open particles [210] or compact structures [211]. In addition, enhanced high-order resonances [212], Fano resonances with two coupled LSP particles [213] or frequency selective propagation of spoof surface plasmons in particle chains [214] have also been demonstrated. On the other hand, the magnetic modes supported by ultrathin spiral particles have also been extensively studied. High-order magnetic LSPs were demonstrated in Reference [215] and the modes supported by the complementary structure were also studied by the same authors [216, 217]. The magnetic coupling among stacked spiral particles was studied in Reference [218]. In addition, spoof LSPs have also been employed to design ultra-compact integrated photonic circuits in microwave and THz frequencies. In particular, in References [219, 220], the spoof-LSPs sustained by ultrathin spoof LSPs structures were coupled to the conformal plasmons propagating along a spoof CSP waveguide. All these works demonstrate the potential of spoof surface plasmons for applications in ultra compact integrated photonic circuitry in the microwave and THz frequency ranges of the spectrum.

6 Conclusions and Outlook

In this Element we have reviewed the fundamentals of spoof SPPs and plasmonic metamaterials based on them. We have discussed how spoof SPPs enable the existence of confined EM modes on metal surfaces at low frequencies (Section 1). Their geometrical origin, as well as their properties, can be described with the coupled mode formalism described in Section 2. Spoof SPPs emerge in flat geometries such as surfaces textured with periodic arrays of grooves or dimples, or slabs perforated with periodic arrays of slits (Section 3). On the other hand, spoof SPP waveguides can be created by periodically corrugating cylindrical wires as well as by corrugating channel waveguides or wedges on flat surfaces (Section 4). Interestingly, spoof SPPs can also be supported on planar platforms made of ultrathin films that can be curved and bent. These are the so-called CSPs (Section 4.6), which have recently received a great deal of attention. Finally, we have also discussed how textured metal particles support the equivalent of the LSP modes that appear in particle plasmonics, with the added ingredient of a magnetic localized mode that exists even for purely metallic structures (Section 5). In the remainder of this Element we give a brief outlook by summarizing some recent applications of spoof SPPs and advances in the field.

We start by discussing the applications of spoof SPPs in THz sensing. THz radiation is nonionizing (it does not damage biological samples usually sensitive to visible light), and many complex molecules have vibrational and rotational modes in this regime of the EM spectrum. This makes THz waves ideal candidates for biological and environmental sensing applications. However, the widespread use of THz spectroscopy in sensing technology is greatly hindered by the low power of sources and poor sensitivity of detectors. These limitations can be circumvented by exploiting the strongly confined EM fields provided by spoof plasmon meta-materials [221, 222]. Through the field enhancement provided by spoof SPP modes, light–matter interactions can be strengthened, which allows for the detection of small changes in the dielectric environment of the THz metamaterial.

Figure 6.1(a) renders the schematics of a spoof plasmon THz sensor [221]. The upper panel shows the metamaterial surface, comprising a periodic array of grooves carved on a thin gold layer deposited on top of a photoresistor and a glass substrate. As depicted in the lower panel, the liquid specimen to be sensed is placed on top of the metamaterial surface, and the whole system is illuminated via prism coupling. A conventional Otto prism (total reflection) configuration for phase-matching allows the evanescent excitation of spoof SPPs. The THz reflectivity spectrum of the sample is then collected in the far field, which inherits the spoof plasmon sensitivity to the refractive index of the specimen.

Figure 6.1(b) plots the experimental spoof SPP dispersion relation measured for the metamaterial surface in panel (a) covered by different fluids [222]. The refractive index of all the substances is between $n = 1$ and $n = 1.85$. We can observe that, as discussed above, a larger refractive index environment increases the effective permittivity of the groove array. Therefore, the binding of the surface EM modes increases and the dispersion band red-shifts. Importantly, the asymptotic region of the band (where the group velocity vanishes) is strongly dependent on the filling n. This fact is

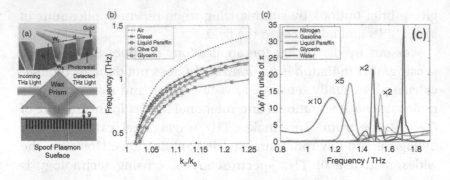

Figure 6.1 (a) Left panel shows a sketch of a spoof plasmon metamaterial device for THz sensing. The pitch of the array is $d = 60\mu$m, height and width of the grooves are $h = 27\mu$m, and $w_t = 37\mu$m (top) and $w_t = 25\mu$m (bottom). Right panel: Schematic of the Otto prism setup used for detection. The liquid sample is placed on top of the spoof plasmon metamaterial. (b) Experimental dispersion relation of the spoof plasmons modes supported by the metamaterial in panel (a) for different fluids filling the grooves. (c) Absolute gradient of phase change spectra for various fluids. Reproduced from [221].

exploited for sensing purposes. Contrary to optical signals, the amplitude and phase of THz waves can be measured independently. The spectral crossing of this asymptotic SPP region leads to a very sharp reflectivity phase gradient.

A figure of merit (FOM) can be defined to assess the sensing performance of the spoof plasmon metamaterial, $FOM = \frac{f_\phi}{\Delta f_\phi}$, where $f_\phi = \Delta\phi'$ is the maximum absolute gradient of phase change and Δf_ϕ is the spectral width of this maximum. Figure 6.1(c) plots phase gradient reflectivity spectra obtained for the THz sensor in panel (a) and for various fluids. It exhibits a phase change maximum for all the substances considered. The FOMs obtained range from 170 (nitrogen) to 7 (water), where the low sensitivity for water is caused by its higher absorption losses in the THz range. These values are significantly higher than those corresponding to amplitude measurements, which proves that the access to phase information in THz spoof SPP metamaterials opens new ways for high precision refractive index spectroscopic sensing of biological fluid samples. Alternatively, spoof THz waveguides can also be used for particle-in-liquid sensing in a microfluidic platform [235].

On the other hand, another very active area of research is the design of spoof SPP based integrated devices and circuits. This was originally motivated by the development of CSPs [94] and spoof LSPs [189]. As we have discussed, these can be supported in ultrathin metal films, which provide a natural platform for the design of integrated functional devices due to their planar character. CSPs offer the possibility of ultrathin waveguides that can be freely bent in a low-loss and broadband manner. This has been utilized to design devices such as bends, beam splitters or filters [94, 173, 223]. In addition, when used in combination with spoof LSPs, CSPs yield great flexibility. Coupling the subwavelength discrete resonances of LSPs with the continuous CSP modes propagating on ultrathin waveguides allows both for a flexible control of the CSP transmission and an efficient excitation means of the LSPs [219, 220]. These kinds of systems are a first step towards integrated circuits and devices at microwave and THz frequencies.

Moreover, active elements based on spoof SPPs have also been proposed and experimentally realized. As opposed to optical frequencies, active devices in microwave and THz frequencies can be realized by taking advantage of the active semiconductor chips available at those frequency ranges. Spoof SPP switches [224–226], amplifiers [179] and frequency mixers [185] have all been demonstrated in this way.

Figure 6.2(a) shows the design of an amplifier [179]. The device uses two microstrip lines as input and output ports, which provide a compact transition in and out of the CSP mode. The CSP mode is supported by a waveguide formed of two ultrathin comb-shaped metal strips placed on top of each other in an antisymmetric manner (see the lower inset of panel (c)). This configuration allows for integration of the semiconductor chip due to the double line structure. In addition, it provides great subwavelength confinement due to the mutual coupling between both lines, which decreases the effective

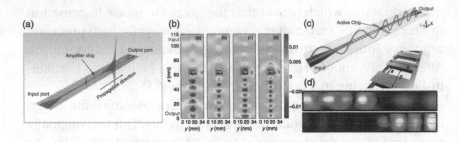

Figure 6.2 (a) Spoof SPPs amplifier composed of microstrip lines as input and output ports, a double-sided CSPs waveguide and a subwavelength semiconductor chip. (b) Measured near-field distribution at 14 GHz, 16 GHz, 18 GHz and 20 GHz (respectively, from left to right) showing how the spoof SPPs are significantly amplified by passing through the subwavelength amplifier. Reproduced from [179]. (c) Second harmonic generation of spoof SPPs through a subwavelength active chip integrated on a plasmonic waveguides connected to microstrip lines. (d) Measured near-field distribution at the input frequency, 8 GHz (top panel), and at the second harmonic, 16 GHz (bottom panel) Reproduced from [185].

plasma frequency. Then, a commercially available low-loss power amplifier is used to amplify the spoof SPPs traveling through the waveguide. As illustrated in panel (b), spoof SPPs are efficiently amplified between 6 and 20 GHz with gain around 20 dB.

On the other hand, a frequency mixer device at microwave frequencies is shown in Figure 6.2(c) [185]. The second harmonics of spoof SPPs were efficiently generated using a subwavelength nonlinear active chip integrated on double sided plasmonic waveguides with the same structure as for the amplifier (see the lower inset of panel (c)). The device is composed of two such waveguides with different corrugation depths and with a field effect transistor positioned at their junction (see upper sketch in panel (c)). As shown in panel (d), an efficient generation of the second harmonic was observed in this unidirectional device for input frequencies between 5 and 10 GHz. Integrating passive and active devices such as this mixer and the amplifier holds great potential for the realization of functional circuits at low frequencies.

Finally, recent advances also include the design of high-efficiency meta-couplers from propagating waves to spoof SPPs [227] based on gradient index metasurfaces [228, 229], or the study of structures with glide symmetry [230, 231]. Furthermore, ideas from the emergent area of topological photonics have also been applied to spoof SPPs. In Reference [232], photonic topological edge states which are robust against imperfections in fabrication or defects have been experimentally shown in a spoof SPPs platform where domino waveguides were arranged in rings, with the rings forming a square lattice. This way, a Floquet topological insulator was designed for spoof SPPs, and its topological protection was experimentally verified.

Notation

1D	One-dimensional
2D	Two-dimensional
3D	Three-dimensional
ACS	Absorption cross section
CPP	Channel plasmon polariton
CSP	Conformal surface plasmon
DP	Domino plasmon
ECS	Extinction cross section
EM	Electromagnetic
IR	Infrared
FDTD	Finite Difference in the Time Domain
FEM	Finite Element Method
FIT	Finite Integration Technique
GHz	Gigahertz
LSP	Localized surface plasmon
PEC	Perfect electrical conductor
SCS	Scattering cross section
SP	Surface plasmon
SPP	Surface plasmon polariton
THz	Terahertz
WPP	Wedge plasmon polariton

References

[1] Smith DR, Pendry JB, Wiltshire MCK. Metamaterials and negative refractive index. *Science*, 2004;305(5685):788–92.

[2] Pendry JB. Photonics: metamaterials in the sunshine. *Nature Materials*. 2006;5(8):599–600.

[3] Shelby RA, Smith DR, Schultz S. Experimental verification of a negative index of refraction. *Science*, 2001;292(5514):77–9.

[4] Schurig D, Mock JJ, Justice BJ, et al. Metamaterial electromagnetic cloak at microwave frequencies. *Science*, 2006;314:977.

[5] Cui, TJ, Smith, DR, Liu, RP. Metamaterials: Theory, Design and Applications. 1st ed. Springer, 2009.

[6] Liu Y, Zhang X. Metamaterials: a new frontier of science and technology. *Chemical Society Reviews*, 2011;40(5):2494–507.

[7] Raether H. *Surface Plasmons on Smooth and Rough Surfaces and on Gratings.* Springer, 1988.

[8] Maier SA. *Plasmonics Fundamentals and Applications.* Boston, MA: Springer, 2007.

[9] Barnes WL, Dereux A, Ebbesen TW. Surface plasmon subwavelength optics. *Nature*, 2003;424(6950):824–30.

[10] Ozbay E. Plasmonics: merging photonics and electronics at nanoscale dimensions. *Science*, 2006 Jan;311(5758):189–93.

[11] Gramotnev DK, Bozhevolnyi SI. Plasmonics beyond the diffraction limit. *Nature Photonics*, 2010;4(2):83–91.

[12] Zhang S, Fan W, Minhas B, Frauenglass A, Malloy K, Brueck S. Midinfrared resonant magnetic nanostructures exhibiting a negative permeability. *Physical Review Letters*, 2005;94(3):037402.

[13] Zhang S, Fan W, Panoiu NC, Malloy KJ, Osgood RM, Brueck SRJ. Experimental demonstration of near-infrared negative-index metamaterials. *Physical Review Letters*, 2005;95(13):137404.

[14] Pendry JB, Martín-Moreno L, García-Vidal FJ. Mimicking surface plasmons with structured surfaces. *Science*, 2004;305(5685):847–8.

[15] García-Vidal FJ, Martín-Moreno L, Pendry JB. Surfaces with holes in them: new plasmonic metamaterials. *Journal of Optics A: Pure and Applied Optics*. 2005;7(2):S97–S101.

[16] Shalaev VM. Optical negative-index metamaterials. *Nature Photonics*. 2007;1(1):41–8.

[17] Cai WS, Shalaev VM. *Optical Metamaterials: Fundamentals and Applications*. 1st ed. New York, NY: Springer, 2009.

[18] Wegener M, Linden S. Shaping optical space with metamaterials feature. *Physics Today*. 2010;63:32–6.

[19] Pendry JB, Holden AJ, Stewart WJ, Youngs I. Extremely low frequency plasmons in metallic mesostructures. *Physical Review Letters*. 1996;76(25):4773–6.

[20] Pendry JB, Holden AJ, Robbins DJ, Stewart WJ. Magnetism from conductors and enhanced nonlinear phenomena. *IEEE Transactions on Microwave Theory and Techniques*, 1999;47(11):2075–2084.

[21] Wiltshire MCK, Pendry JB, Young IR, Larkman DJ, Gilderdale DJ, Hajnal JV. Microstructured magnetic materials for RF flux guides in magnetic resonance imaging. *Science*, 2001;291(5505):849.

[22] Soukoulis CM, Linden S, Wegener M. Negative refractive index at optical wavelengths. *Science*, 2007;315(5808):47–9.

[23] Lezec HJ, Dionne JA, Atwater HA. Negative refraction at visible frequencies. *Science*, 2007;316(5823):430–2.

[24] Yao J, Liu Z, Liu Y, et al. Optical negative refraction in bulk metamaterials of nanowires. *Science*, 2008;321(5891):930.

[25] Valentine J, Zhang S, Zentgraf T, et al. Three-dimensional optical metamaterial with a negative refractive index. *Nature*, 2008;455(7211):376–9.

[26] Fang N, Lee H, Sun C, Zhang X. Subdiffraction-limited optical imaging with a silver superlens. *Science*, 2005;308(5721):534–7.

[27] Taubner T, Korobkin D, Urzhumov Y, Shvets G, Hillenbrand R. Near-field microscopy through a SiC superlens. *Science*, 2006;313 (5793):1595.

[28] Zhang X, Liu Z. Superlenses to overcome the diffraction limit. *Nature Materials*, 2008;7(6):435–41.

[29] Zhang S, Park YS, Li J, Lu X, Zhang W, Zhang X. Negative Refractive Index in Chiral Metamaterials. *Physical Review Letters*, 2009;102(2):023901.

[30] Gansel JK, Thiel M, Rill MS, et al. Gold helix photonic metamaterial as broadband circular polarizer. *Science*, 2009;325(5947):1513–5.

[31] Kaelberer T, Fedotov VA, Papasimakis N, Tsai DP, Zheludev NI. Toroidal dipolar response in a metamaterial. *Science*, 2010;330 (6010):1510–12.

[32] Kabashin AV, Evans P, Pastkovsky S, et al. Plasmonic nanorod metamaterials for biosensing. *Nature Materials*, 2009;8(11):867–71.

[33] Wu C, Khanikaev AB, Adato R, et al. Fano-resonant asymmetric metamaterials for ultrasensitive spectroscopy and identification of molecular monolayers. *Nature Materials*, 2011;11(1):69–75.

[34] Sreekanth KV, Alapan Y, ElKabbash M, et al. Extreme sensitivity biosensing platform based on hyperbolic metamaterials. *Nature Materials*, 2016;15(March):4–11.

[35] Soukoulis CM, Wegener M. Past achievements and future challenges in the development of three-dimensional photonic metamaterials. *Nature Photonics*, 2011;5(9):523.

[36] Hess O, Pendry JB, Maier SA, Oulton RF, Hamm JM, Tsakmakidis KL. Active nanoplasmonic metamaterials. *Nature Materials*, 2012;11(7):573–84.

[37] Neira AD, Olivier N, Nasir ME, Dickson W, Wurtz GA, Zayats AV. Eliminating material constraints for nonlinearity with plasmonic metamaterials. *Nature Communications*. 2015;6:7757.

[38] Meinzer N, Barnes WL, Hooper IR. Plasmonic meta-atoms and metasurfaces. *Nature Photonics*, 2014;8(12):009–98.

[39] Kildishev AV, Boltasseva A, Shalaev VM. Planar photonics with metasurfaces. *Science*, 2013;339(6125):1232009.

[40] Ni X, Emani NK, Kildishev AV, Boltasseva A, Shalaev VM. Broadband light bending with plasmonic nanoantennas. *Science*, 2012;335 (6067):427.

[41] Yu N, Genevet P, Kats MA, et al. Light propagation with phase discontinuities: generalized laws of reflection and refraction, *Science*, 2011;334(6054):333–7.

[42] Ding F, Wang Z, He S, Shalaev V, Kildishev A. Broadband high-efficiency half-wave plate: a super-cell based plasmonic metasurface approach. *ACS Nano*, 2015;9(4):4111–19.

[43] Yin X, Ye Z, Rho J, Wang Y, Zhang X. Photonic Spin Hall Effect at Metasurfaces. *Science*, 2013;339(6126):1405–7.

[44] Khorasaninejad M, Chen WT, Devlin RC, Oh J, Zhu AY, Capasso F. Metalenses at visible wavelengths: diffraction-limited focusing and subwavelength resolution imaging. *Science*, 2016;352 (6290):1190–4.

[45] Ramakrishna SA. Physics of negative refractive index materials. *Reports on Progress in Physics*, 2005;68(2):449–521.

[46] Murray WA, Barnes WL. Plasmonic materials. Advanced Materials, 2007;19(22):3771–82.

[47] Maier SA, Atwater HA. Plasmonics: Localization and guiding of electromagnetic energy in metal/dielectric structures. *Journal of Applied Physics*, 2005;98(1):011101.

[48] Schuck PJ, Fromm DP, Sundaramurthy A, Kino GS, Moerner WE. Improving the mismatch between light and nanoscale objects with gold bowtie nanoantennas. *Physical Review Letters*, 2005;94 (1):017402.

[49] Mühlschlegel P, Eisler HJ, Martin OJF, Hecht B, Pohl DW. Resonant optical antennas. *Science*, 2005;308(5728):1607–9.

[50] Anger P, Bharadwaj P, Novotny L. Enhancement and quenching of single-molecule fluorescence, *Physical Review Letters*, 2006;96 (11):113002(1–4).

[51] Kühn S, Hakanson U, Rogobete L, Sandoghdar V. Enhancement of single-molecule fluorescence using a gold nanoparticle as an optical nanoantenna. *Physical Review Letters*, 2006;97(1):017402(1–4).

[52] Novotny L. Effective wavelength scaling for optical antennas. *Physical Review Letters*, 2007;98(26):266802.

[53] Ghenuche P, Cherukulappurath S, Taminiau TH, van Hulst NF, Quidant R. Spectroscopic mode mapping of resonant plasmon nanoantennas. *Physical Review Letters*, 2008;101(11):116805.

[54] Bryant GW, García de Abajo FJ, Aizpurua J. Mapping the plasmon resonances of metallic nanoantennas. *Nano Letters*, 2008;8(2):631–6.

[55] Kinkhabwala A, Yu Z, Fan S, Avlasevich Y, Müllen K, Moerner WE. Large single-molecule fluorescence enhancements produced by a bowtie nanoantenna. *Nature Photonics*, 2009;3(11):654–7.

[56] Curto AG, Volpe G, Taminiau TH, Kreuzer MP, Quidant R, van Hulst NF. Unidirectional emission of a quantum dot coupled to a nanoantenna. *Science*, 2010;329(5994):930–3.

[57] Schuller JA, Barnard ES, Cai W, Jun YC, White JS, Brongersma ML. Plasmonics for extreme light concentration and manipulation. *Nature Materials*, 2010;9(3):193–204.

[58] Atwater HA, Polman A. Plasmonics for improved photovoltaic devices. *Nature Materials*, 2010;9(3):865.

[59] Fan JA, Wu C, Bao K, et al. Self-Assembled Plasmonic Nanoparticle Clusters. *Science*, 2010;328(5982):1135–8.

[60] Novotny L, van Hulst NF. Antennas for light. *Nature Photonics*, 2011;5(2):83–90.

[61] Höppener C, Lapin ZJ, Bharadwaj P, Novotny L. Self-similar gold-nanoparticle antennas for a cascaded enhancement of the optical field. *Physical Review Letters*, 2012;109(1):017402.

[62] Rodrigo S, García-Vidal FJ, Martín-Moreno L. Influence of material properties on extraordinary optical transmission through hole arrays. *Physical Review B*, 2008;77(7):075401.

[63] Johnson PB, Christy RW. Optical constants of noble metals. *Physical Review B*, 1972;6(12):4370–9.

[64] Palik E. *Handbook of Optical Constants of Solids*, edited by Edward D. Palik. Academic Press Handbook Series, New York, NY: Academic Press, 1985.

[65] Novotny L, Hetch B. *Principles of Nanooptics*, 1st ed. Cambridge: Cambridge University Press, 2006.

[66] Archambault A, Teperik TV, Marquier F, Greffet JJ. Surface plasmon Fourier optics. *Physical Review B – Condensed Matter and Materials Physics*, 2009;79(19):1–8.

[67] Otto A. Excitation of nonradiative surface plasma waves in silver by the method of frustrated total reflection. *Zeits Phys.*, 1968;216(4):398–410.

[68] Kretschmann E, Raether H. Radiative decay of non-radiative surface plasmons excited by light. *Z Naturforschung, A.*, 1968;23:2135.

[69] Pelton M, Aizpurua J, Bryant G. Metal-nanoparticle plasmonics. *Laser & Photonics Review*. 2008;2(3):136–59.

[70] Giannini V, Fernández-Domínguez AI, Heck SC, Maier SA. Plasmonic nanoantennas: fundamentals and their use in controlling the radiative properties of nanoemitters. *Chemical Reviews*, 2011;111(6):3888–912.

[71] Jackson JD. *Classical Electrodynamics*, 3rd ed. Wiley, 1998.

[72] Zenneck J. Propagation of plane electromagnetic waves along a plane conducting surface. *Ann Phys(Leipzig)*, 1907;23(1):846.

[73] Sommerfeld A. Propagation of electrodynamic waves along a cylindric conductor. *Ann Phys und Chemie*, 1899;67:233.

[74] Gómez-Rivas J, Kuttge M, Bolivar PH, Kurz H, Sánchez-Gil JA. Propagation of Surface Plasmon Polaritons on Semiconductor Gratings. *Phys Rev Lett.*, 2004;93(25):256804.

[75] Hanham SM, Maier SA. Chapter 8 in *Terahertz Plasmonic Surfaces for Sensing*. John Wiley & Sons, Inc., 2013, pp. 243–60.

[76] Gobau G. Surface waves and their application to transmission lines. *J Appl Phys*, 1950;21:1119.

[77] Mills DL, Maradudin AA. Surface corrugation and surface-polariton binding in the infrared frequency range. *Phys Rev B*, 1989;39:1569.

[78] Munk BA. *Frequency Selective Surfaces: Theory and Design*. New York, NY: Wiley, 2000.

[79] Ulrich R, Tacke M. Submilimeter waveguiding on periodic metal structure. *Appl Phys Lett.*, 1973;22:251.

[80] Hibbins AP, Evans BR, Sambles JR. Experimental verification of designer surface plasmons. *Science*, 2005;308(5722):670–2.

[81] Hibbins A, Lockyear M, Hooper I, Sambles J. Waveguide arrays as plasmonic metamaterials: transmission below cutoff. *Physical Review Letters*, 2006;96(7):073904.

[82] Williams CR, Andrews SR, Maier SA, Fernández-Domínguez AI, Martín-Moreno L, García-Vidal FJ. Highly confined guiding of terahertz surface plasmon polaritons on structured metal surfaces. *Nature Photonics*, 2008;2(3):175–9.

[83] Yu N, Wang QJ, Kats MA, Fan JA, Khanna SP, Li L, et al. Designer spoof surface plasmon structures collimate terahertz laser beams. *Nature Materials*, 2010;9(9):730–5.

[84] García de Abajo FJ, Sáenz JJ. Electromagnetic surface modes in structured perfect-conductor surfaces. *Physical Review Letters*, 2005;95(23):233901.

[85] Hendry E, Hibbins AP, Sambles JR. Importance of diffraction in determining the dispersion of designer surface plasmons. *Physical Review B*, 2008;78(23):235426.

[86] Maier SA, Andrews SA, Martín-Moreno L, García-Vidal FJ. Terahertz surface plasmon-polariton propagation and focusing on periodically corrugated metal wires. *Physical Review Letters*, 2006;97(17):176805.

[87] Fernández-Domínguez AI, Moreno E, Martín-Moreno L, García-Vidal FJ. Terahertz wedge plasmon polaritons. *Optics Letters*, 2009;34(13):2063–5.

[88] Fernández-Domínguez AI, Moreno E, Martín-Moreno L, García-Vidal FJ. Guiding terahertz waves along subwavelength channels. *Physical Review B*, 2009;79(23):233104.

[89] Martín-Cano D, Nesterov ML, Fernández-Domínguez AI, García-Vidal FJ, Martín-Moreno L, Moreno E. Domino plasmons for subwavelength terahertz circuitry. *Optics Express*, 2010;18(2):754–64.

[90] Kats MA, Woolf D, Blanchard R, Yu N, Capasso F. Spoof plasmon analogue of metal-insulator-metal waveguides. *Optics Express*, 2011;19(16):14860–70.

[91] Fernández-Domínguez AI, Williams CR, García-Vidal FJ, Martín-Moreno L, Andrews SR, Maier SA. Terahertz surface plasmon polaritons on a helically grooved wire. *Applied Physics Letters*, 2008;93(14):141109.

[92] Brock EMG, Hendry E, Hibbins AP. Subwavelength lateral confinement of microwave surface waves. *Applied Physics Letters*, 2011;99(5):051108.

[93] Nesterov ML, Martín-Cano D, Fernández-Domínguez AI, Moreno E, Martín-Moreno L, García-Vidal FJ. Geometrically induced modification of surface plasmons in the optical and telecom regimes. *Optics Letters*, 2010;35:423–5.

[94] Shen X, Cui TJ, Martín-Cano D, García-Vidal FJ. Conformal surface plasmons propagating on ultrathin and flexible films. *Proceedings of the National Academy of Sciences*, 2013;110(1):40–5.

[95] Pors A, Moreno E, Martín-Moreno L, Pendry JB, García-Vidal FJ. Localized spoof plasmons arise while texturing closed surfaces. *Physical Review Letters*, 2012;108(22):223905.

[96] Huidobro PA, Moreno E, Martin-Moreno L, Pendry JB, García-Vidal FJ. Magnetic localized surface plasmons supported by metal structures, in *9th International Congress on Advanced Electromagnetic Materials in Microwaves and Optics (METAMATERIALS)*, 2014. pp. 13–15.

[97] Martín-Moreno L, García-Vidal FJ, Lezec HJ, et al. Theory of extraordinary optical transmission through subwavelength hole arrays. *Phys Rev Lett.*, 2001;86:1114.

[98] Bravo-Abad J, García-Vidal FJ, Martín-Moreno L. Resonant transmission of light through finite chains of subwavelength holes in a metallic film. *Phys Rev Lett.*, 2004;93:227401.

[99] Mary A, Rodrigo SG, García-Vidal FJ, Martín-Moreno L. Theory of negative-refractive-index response of double-fishnet structures. *Phys Rev Lett.*, 2008;101:103902.

[100] Qiu M. Photonic band structures for surface waves on structured metal surfaces. *Opt. Express*, 2005;13:7583.

[101] Fernández-Domínguez AI, Martín-Moreno L, García-Vidal FJ. Chapter 7, in Maradudin AA, editor, *Surface Electromagnetic*

Waves on Structured Perfectly Conducting Surfaces. Cambridge: Cambridge University Press, 2011, pp. 232–65.

[102] Morse PM, Feshbach H. *Methods of Theoretical Physics*. New York, NY: McGraw-Hill, 1953.

[103] Roberts A. Electromagnetic theory of diffraction by a circular aperture in a thick, perfectly conducting screen. *J Opt Soc Am A.*, 1987;4:1970.

[104] Wood JJ, Tomlinson LA, Hess O, Maier SA, Fernández-Dominguez AI. Spoof plasmon polaritons in slanted geometries. *Phys Rev B*, 2012;85:075441.

[105] Kim SH, Oh SS, Kim KJ, et al. Subwavelength localization and toroidal dipole moment of spoof surface plasmon polaritons. *Physical Review B – Condensed Matter and Materials Physics*, 2015;91(3):1–9.

[106] Gao Z, Gao F, Zhang B. Guiding, bending, and splitting of coupled defect surface modes in a surface-wave photonic crystal. *Applied Physics Letters*, 2016;108(4):9–14.

[107] Woolf D, Kats Ma, Capasso F. Spoof surface plasmon waveguide forces. *Optics Letters*. 2014;39(3):517–20.

[108] Rodriguez AW, Hui PC, Woolf DP, Johnson SG, Lončar M, Capasso F. Classical and fluctuation-induced electromagnetic interactions in micron-scale systems: designer bonding, antibonding, and Casimir forces. Annalen der Physik, 2015;527(1–2):45–80.

[109] Davids PS, Intravaia F, Dalvit DaR. Spoof polariton enhanced modal density of states in planar nanostructured metallic cavities. *Optics Express*, 2014;22(10):12424–37.

[110] Dai J, Dyakov SA, Yan M. Enhanced near-field radiative heat transfer between corrugated metal plates: Role of spoof surface plasmon polaritons. *Physical Review B*, 2015;92(3):035419.

[111] Ooi K, Okada T, Tanaka K. Mimicking electromagnetically induced transparency by spoof surface plasmons. *Phys Rev B*, 2011;84 (11):115405.

[112] Shen JT, Catrysse PB, Fan S. Mechanism for designing metallic metamaterials with a high index of refraction. *Phys Rev Lett*, 2005;94:197401.

[113] Shin J, Shen JT, Catrysse PB, Fan S. Cut-through metal slit array as an anisotropic metamaterial film. *IEEE J Selected Topics in Quant Elec.*, 2006;12:1116.

[114] Shin YM, So JK, Won JH, Park GS. Frequency-dependent refractive index of one-dimensionally structured thick metal film. *Appl Phys Lett.*, 2007;91:031102.

[115] Zhang XF, Shen LF, Ran LX. Low-frequency surface plasmon polaritons propagating along a metal film with periodic cut-through slits in symmetric and asymmetric environments. *J Appl Phys.*, 2009;105:013704.

[116] Economou EN. Surface Plasmons in Thin Films. *Phys Rev.*, 1969;182:539.

[117] Shen L, Chen X, Yang TJ. Terahertz surface plasmon polaritons on periodically corrugated metal surfaces. *Optics Express*, 2008;16:3326.

[118] Collin S, Sauvan C, Billaudeau C, et al. Surface modes on nanostructured metallic surfaces. *Phys Rev B*, 2009;79:165405.

[119] Hibbins AP, Hendry E, Lockyear MJ, Sambles JR. Prism coupling to 'designer' surface plasmons. *Optics Express*, 2008;16:20441.

[120] Ferguson BF, Zhang XC. Materials for terahertz science and technology. *Nature Materials*, 2002;1:26.

[121] Tonouchi M. Cutting-edge terahertz technology. *Nature Photonics*, 2007;1:97–105.

[122] Agrawal A, Vardeny ZV, Nahata A. Engineering the dielectric function of plasmonic lattices. *Optics Express*, 2008;16:9601.

[123] Zhu W, Agrawal A, Nahata A. Planar plasmonic terahertz guided-wave devices. *Optics Express*, 2008;16:6216.

[124] Lan YC, Chern RL. Surface plasmon-like modes on structured perfectly conducting surfaces. *Optics Express*, 2006;14:11339.

[125] Ruan ZC, Qiu M. Slow electromagnetic wave guided in subwavelength regions along one-dimensional periodically structured metal surface. *Appl Phys Lett.*, 2007;90:201906.

[126] Lockyear MJ, Hibbins AP, Sambles JR. Microwave surface-plasmon-like modes on thin metamaterials. *Phys Rev Lett.*, 2009;102:073901.

[127] Navarro-Cía M, Beruete M, Agrafiotis S, Falcone F, Sorolla M, Maier SA. Broadband spoof plasmons and subwavelength electromagnetic energy confinement on ultrathin metafilms. *Optics Express*, 2009;17:18184.

[128] Williams CR, Misra M, Andrews SR, et al. Dual band terahertz waveguidng on a planar metal surface patterned with annular holes. *Appl Phys Lett.*, 2010;96:011101.

[129] Gan Q, Fu Z, Ding YJ, Bartoli FJ. Ultrawide-bandwidth slow-light system based on THz plasmonic graded metallic grating structures. *Phys Rev Lett.*, 2008;100:256803.

[130] Maier SA, Andrews SR. Terahertz pulse propagation using plasmon-polariton-like surface modes on structured conductive surfaces. *Appl Phys Lett.*, 2006;88:251120.

[131] Juluri BK, Lin SCS, Walker TR, Jensen L, Huang TJ. Propagation of designer surface plasmons in structured conductor surfaces with parabolic gradient index. *Optics Express*, 2009;17:2997.

[132] Song K, Mazumder P. Active terahertz spoof surface plasmon polariton switch comprising the perfect conductor metamaterial. *IEEE Trans Elec Dev.*, 2009;56:2792.

[133] Wang K, Mittleman DM. Metal wires for terahertz waveguiding. *Nature*, 2004;432:376.

[134] Jeon TI, Zhang J, Grischkowsky D. THz Sommerfeld wave propagation on a single metal wire. *Appl Phys Lett.*, 2005;86:161904.

[135] Piefke G. The transmission characteristics of a corrugated wire. *IRE Trans Antennas Propag.*, 1959;7:183.

[136] Fernández-Domínguez AI, Martín-Moreno L, García-Vidal FJ, Andrews SR, Maier SA. Spoof surface plasmon polariton modes propagating along periodically corrugated wires. *IEEE J Sel Top Quant Elect.*, 2008;14:1515.

[137] Chen Y, Song Z, Li Y, et al. Effective surface plasmon polaritons on the metal wire with arrays of subwavelength grooves. *Optics Express*, 2006;14:13021.

[138] Arfken GB, Weber HJ. *Mathematical Methods for Physicists*, 5th ed. London: Harcourt Academic Press, 2001.

[139] Stockman M. Nanofocusing of optical energy in tapered plasmonic waveguides. *Physical Review Letters*, 2004;93(13):1–4.

[140] Ruting F, Fernández-Dominguez AI, Martín-Moreno L, García-Vidal FJ. Subwavelength chiral surface plasmons that carry tuneable orbital angular momentum. *Phys Rev B*, 2012;86:075437.

[141] Fernández-Domínguez AI, Williams CR, Martín-Moreno L, García-Vidal FJ, Andrews SR, Maier SA. Terahertz surface plasmon polaritons on a helically grooved wire. *Apl Phys Lett.*, 2008;93:141109.

[142] Pendry JB. A chiral route to negative refraction. *Science*, 2004;306 (5700):1353–5.

[143] Crepeau PJ. Consequences of Symmetry in Periodic Structures. *Proc IEEE.*, 1964;52:33.

[144] Novikov IV, Maradudin AA. Channel polaritons. *Phys Rev B*, 2002;66:035403.

[145] Bozhevolnyi SI, Volkov VS, Devaux E, Laluet JY, Ebbesen TW. Channel plasmon subwavelength waveguide components including interferometers and ring resonators. *Nature*, 2006;440:508.

[146] Gao Z, Shen L, Zheng X. Highly-confined guiding of terahertz waves along subwavelength grooves. *IEEE Photonics Technology Letters*, 2012;24(15):1343-5.

[147] Jiang T, Shen L, Wu JJ, Yang TJ, Ruan Z, Ran L. Realization of tightly confined channel plasmon polaritons at low frequencies. *Applied Physics Letters*, 2011;99(26):261103.

[148] Zhou YJ, Jiang Q, Cui TJ. Bidirectional bending splitter of designer surface plasmons. *Applied Physics Letters*, 2011;99(11):111904.

[149] Li X, Jiang T, Shen L, Deng X. Subwavelength guiding of channel plasmon polaritons by textured metallic grooves at telecom wavelengths. *Applied Physics Letters*, 2013;102(3):031606.

[150] Fernández-Domínguez AI, Moreno E, Martín-Moreno L, García-Vidal FJ. Guiding terahertz waves along subwavelength channels. *Phys Rev B*, 2009;79:233104.

[151] Moreno E, Garcia-Vidal FJ, Rodrigo SG, Martin-Moreno L, Bozhevolnyi SI. Channel plasmon-polaritons: modal shape, dispersion, and losses. *Opt Lett.*, 2006 Dec;31(23):3447-3449.

[152] Fernández-Domínguez AI, Moreno E, Martín-Moreno L, García-Vidal FJ. Terahertz wedge plasmon polaritons. *Opt Lett.*, 2009 Jul;34(13):2063-2065.

[153] Pile DFP, Gramotnev DK. Channel plasmon-polariton in a triangular groove on a metal surface. *Opt Lett.*, 2004;29(10):1069.

[154] Moreno E, Rodrigo SG, Bozhevolnyi SI, Martín-Moreno L, García-Vidal FJ. Guiding and focusing of electromagnetic fields with wedge plasmon polaritons. *Phys Rev Lett.*, 2008;100(2):023901.

[155] Gao Z, Zhang X, Shen L. Wedge mode of spoof surface plasmon polaritons at terahertz frequencies. *Journal of Applied Physics*, 2010;108(11):113104.

[156] Zhao W, Eldaiki OM, Yang R, Lu Z. Deep subwavelength waveguiding and focusing based on designer surface plasmons. *Optics Express*, 2010;18(20):21498-21503.

[157] Ma YG, Lan L, Zhong SM, Ong CK. Experimental demonstration of subwavelength domino plasmon devices for compact high-frequency circuit. *Optics Express*, 2011;19(22):21189.

[158] Kumar G, Li S, Jadidi MM, Murphy TE. Terahertz surface plasmon waveguide based on a one-dimensional array of silicon pillars. *New Journal of Physics*, 2013;15(8).

[159] Pandey S, Gupta B, Nahata A. Terahertz plasmonic waveguides created via 3D printing. *Optics Express*, 2013;21(21):24422.

[160] Martín-Cano D, Quevedo-Teruel O, Moreno E, Martín-Moreno L, García-Vidal FJ. Waveguided spoof surface plasmons with deep-subwavelength lateral confinement. *Optics Letters*, 2011;36 (23):4635-7.

[161] Gupta B, Pandey S, Nahata A. Plasmonic waveguides based on symmetric and asymmetric T-shaped structures. *Optics Express*, 2014;22(3):2868.

[162] Shen L, Chen X, Zhang X, Agarwal K. Guiding terahertz waves by a single row of periodic holes on a planar metal surface. *Plasmonics*, 2011;6(2):301-5.

[163] Hooper IR, Tremain B, Dockrey JA, Hibbins AP. Massively sub-wavelength guiding of electromagnetic waves. *Scientific Reports*, 2014;4:7495.

[164] Quesada R, Martín-Cano D, García-Vidal FJ, Bravo-Abad J. Deep-subwavelength negative-index waveguiding enabled by coupled conformal surface plasmons. *Optics Letters*, 2014;39(10):2990.

[165] Liu L, Li Z, Xu B, Ning P, Chen C, Xu J, et al. Dual-band trapping of spoof surface plasmon polaritons and negative group velocity realization through microstrip line with gradient holes. *Applied Physics Letters*, 2015;107(20).

[166] Liu X, Feng Y, Chen K, Zhu B, Zhao J, Jiang T. Planar surface plasmonic waveguide devices based on symmetric corrugated thin film structures. *Optics Express*, 2014;22(17):20107.

[167] Gao X, Hui Shi J, Shen X, et al. Ultrathin dual-band surface plasmonic polariton waveguide and frequency splitter in microwave frequencies. *Applied Physics Letters*, 2013;102(15):1-5.

[168] Liu X, Feng Y, Zhu B, Zhao J, Jiang T. High-order modes of spoof surface plasmonic wave transmission on thin metal film structure. *Optics Express*, 2013;21(25):31155-65.

[169] Ma HF, Shen X, Cheng Q, Jiang WX, Cui TJ. Broadband and high-efficiency conversion from guided waves to spoof surface plasmon polaritons. *Laser and Photonics Reviews*, 2014;8(1):146-51.

[170] Gao X, Zhou L, Yu XY, et al. Ultra-wideband surface plasmonic Y-splitter. *Optics Express*, 2015;23(18):23270.

[171] Han Z, Zhang Y, Bozhevolnyi SI. Spoof surface plasmon-based stripe antennas with extreme field enhancement in the terahertz regime. *Optics Letters*. 2015;40(11):2533–6.

[172] Yin JY, Ren J, Zhang HC, Pan BC, Cui TJ. Broadband frequency-selective spoof surface plasmon polaritons on ultrathin metallic structure. *Scientific Reports*, 2015;5:8165.

[173] Gao X, Zhou L, Liao Z, Ma HF, Cui TJ. An ultra-wideband surface plasmonic filter in microwave frequency. *Applied Physics Letters*, 2014;104(19):17–22.

[174] Zhang Q, Zhang HC, Wu H, Cui TJ. A Hybrid Circuit for Spoof Surface Plasmons and Spatial Waveguide Modes to Reach Controllable Band-Pass Filters. *Scientific Reports*, 2015;5(4):16531.

[175] Zhang Q, Zhang HC, Yin JY, Pan BC, Cui TJ. A series of compact rejection filters based on the interaction between spoof SPPs and CSRRs. *Scientific Reports*. 2016;6(4):28256.

[176] Xu J, Li Z, Liu L, et al. Low-pass plasmonic filter and its miniaturization based on spoof surface plasmon polaritons. *Optics Communications*. 2016;372:155–9.

[177] Yang Y, Chen H, Xiao S, Mortensen NA, Zhang J. Ultrathin 90-degree sharp bends for spoof surface plasmon polaritons. *Optics Express*, 2015;23(15):19074.

[178] Liang Y, Yu H, Zhang HC, Yang C, Cui TJ. On-chip sub-terahertz surface plasmon polariton transmission lines in CMOS. *Scientific Reports*, 2015;5:14853.

[179] Zhang HC, Liu S, Shen X, Chen LH, Li L, Cui TJ. Broadband amplification of spoof surface plasmon polaritons at microwave frequencies. *Laser and Photonics Reviews*, 2015;9(1):83–90.

[180] Yang Y, Shen X, Zhao P, Zhang HC, Cui TJ. Trapping surface plasmon polaritons on ultrathin corrugated metallic strips in microwave frequencies. *Optics Express*, 2015;23(6):7031.

[181] Zhang W, Zhu G, Sun L, Lin F. Trapping of surface plasmon wave through gradient corrugated strip with underlayer ground and manipulating its propagation. *Applied Physics Letters*, 2015;106 (2):17–22.

[182] Yin JY, Ren J, Zhang HC, Zhang Q, Cui TJ. Capacitive-coupled series spoof surface plasmon polaritons. *Scientific Reports*, 2016;6:24605.

[183] Pan BC, Zhao J, Liao Z, Zhang HC, Cui TJ. Multi-layer topological transmissions of spoof surface plasmon polaritons. *Scientific Reports*, 2016;6:22702.

[184] Li Y, Zhang J, Qu S, Wang J, Feng M, Wang J. K-dispersion engineering of spoof surface plasmon polaritons for beam steering. *Optics Express*, 2016;24(2):2569–2571.

[185] Zhang HC, Fan Y, Guo J, Fu X, Cui TJ. Second-harmonic generation of spoof surface plasmon polaritons using nonlinear plasmonic metamaterials. *ACS Photonics*, 2016;3(1):139–146.

[186] Zhang HC, Cui TJ, Zhang Q, Fan Y, Fu X. Breaking the challenge of signal integrity using time-domain spoof surface plasmon polaritons. *ACS Photonics*, 2015;2(9):1333–1340.

[187] Xiang H, Meng Y, Zhang Q, Qin FF, Xiao JJ, Han D, et al. Spoof surface plasmon polaritons on ultrathin metal strips with tapered grooves. *Optics Communications*, 2015;356:59–63.

[188] Yang BJ, Zhou YJ. Compact broadband slow wave system based on spoof plasmonic THz waveguide with meander grooves. *Optics Communications*, 2015;356:336–342.

[189] Huidobro PA, Shen X, Cuerda J, Moreno E, Martín-Moreno L, García-Vidal FJ, et al. Magnetic localized surface plasmons. *Physical Review X*, 2014;4(2):021003.

[190] Harvey AF. Periodic and guiding structures at microwave frequencies. *IRE Transactions on microwave theory and techniques*, 1960;8:30–61.

[191] Kildal PS. Artificially soft and hard surfaces in electromagnetics. *IEEE Transactions on Antennas and Propagation*, 1990;38 (10):1537–1544.

[192] Shen X, Cui TJ. Ultrathin plasmonic metamaterial for spoof localized surface plasmons. *Laser and Photonics Reviews*, 2014;8(1): 137–145.

[193] Liao Z, Luo Y, Fernández-Domínguez AI, Shen X, Maier Sa, Cui TJ. High-order localized spoof surface plasmon resonances and experimental verifications. *Scientific Reports*, 2015;5:9590.

[194] Bohren CF, Huffman DR. *Absorption and Scattering of Light by Small Particles*. John Wiley and Sons, 1983.

[195] García-Etxarri A, Gómez-Medina R, Froufe-Pérez LS, et al. Strong magnetic response of submicron silicon particles in the infrared. *Optics Express*, 2011;19(6):4815–26.

[196] Kuznetsov AI, Miroshnichenko AE, Fu YH, Zhang J, Luk'yanchuk B. Magnetic light. *Scientific Reports*, 2012;2:492.

[197] Dyson JD. The equiangular spiral antenna. *IEEE Transactions on antennas and propagation*, 1959;2:181.

[198] Kaiser JA. The Archimedean two-wire spiral antenna. *IEEE Transactions on antennas and propagation.* 1960;8:312.

[199] Balanis CA. *Antenna Theory: Analysis and Design*, 3rd ed. Wiley-Interscience, 2005.

[200] Baena JD, Marqués R, Medina F, Martel J. Artificial magnetic metamaterial design by using spiral resonators. *Physical Review B*, 2004;69(1):014402.

[201] Bilotti F, Toscano A, Vegni L. Design of spiral and multiple split-ring resonators for the realization of miniaturized metamaterial samples. *IEEE Transactions on Antennas and Propagation*, 2007;55(8):2258–2267.

[202] Zhu X, Liang B, Kan W, Peng Y, Cheng J. Deep-subwavelength-scale directional sensing based on highly localized dipolar mie resonances. *Physical Review Applied*, 2016;5(5):054015.

[203] Ordal MA, Long LL, Bell RJ, et al. Optical properties of the metals Al, Co, Cu, Au, Fe, Pb, Ni, Pd, Pt, Ag, Ti, and W in the infrared and far infrared. *Applied Optics*, 1983;22(7):1099–1120.

[204] CST. *Microwave Studio* (computer software): www.cst.com/pro ducts/cstmws.

[205] Liao Z, Liu S, Ma HF, Li C, Jin B, Cui TJ. Electromagnetically induced transparency metamaterial based on spoof localized surface plasmons at terahertz frequencies. *Scientific Reports*, 2016;6 (4):27596.

[206] Li Z, Xu B, Gu C, Ning P, Liu L, Niu Z, et al. Localized spoof plasmons in closed textured cavities. *Applied Physics Letters*, 2014;104(25):251601.

[207] Xu B, Li Z, Gu C, Ning P, Liu L, Niu Z, et al. Multiband localized spoof plasmons in closed textured cavities. *Appl Opt.*, 2014;53 (30):6950–3.

[208] Yang BJ, Zhou YJ, Xiao QX. Spoof localized surface plasmons in corrugated ring structures excited by microstrip line. *Optics Express*, 2015;23(16):21434.

[209] Zhou YJ, Xiao QX, Jia Yang B. Spoof localized surface plasmons on ultrathin textured MIM ring resonator with enhanced resonances. *Scientific Reports*, 2015;5(September):14819.

[210] Gao Z, Gao F, Xu H, Zhang Y, Zhang B. Localized spoof surface plasmons in textured open metal surfaces. *Optics Letters*, 2016;41 (10):3–6.

[211] Ao DIB, Ajab KHZR, Iang WEIXIJ, Heng QIC, Iao ZHENL. Experimental demonstration of compact spoof localized surface plasmons. *Optics Letters*, 2016;41(23):5418–21.

[212] Gao F, Gao Z, Shi X, Yang Z, Lin X, Zhang B. Dispersion-tunable designer-plasmonic resonator with enhanced high-order resonances. *Optics Express*, 2015;23(5):6896–902.

[213] Xiao QX, Yang BJ, Zhou YJ. Spoof localized surface plasmons and Fano resonances excited by flared slot line. *Journal of Applied Physics*, 2015;118(23):1–6.

[214] Gao Z, Gao F, Shastri KK, Zhang B. Frequency-selective propagation of localized spoof surface plasmons in a graded plasmonic resonator chain. *Scientific Reports*, 2016;6(April):25576.

[215] Gao Z, Gao F, Zhang Y, Shi X, Yang Z, Zhang B. Experimental demonstration of high-order magnetic localized spoof surface plasmons. *Applied Physics Letters*, 2015;107(4):1–5.

[216] Gao Z, Gao F, Zhang Y, Zhang B. Complementary structure for designer localized surface plasmons. *Applied Physics Letters*, 2015;107(19):191103.

[217] Gao Z, Gao F, Zhang B. High-order spoof localized surface plasmons supported on a complementary metallic spiral structure. *Scientific Reports*, 2016;6(April):24447.

[218] Gao Z, Gao F, Zhang Y, Zhang B. Deep-subwavelength magnetic-coupling-dominant interaction among magnetic localized surface plasmons. *Physical Review B*, 2016;93(19):195410.

[219] Shen X, Jun Cui T. Planar plasmonic metamaterial on a thin film with nearly zero thickness. *Applied Physics Letters*, 2013;102 (21):14–18.

[220] Shen X, Pan BC, Zhao J, Luo Y, Cui TJ. A combined system for efficient excitation and capture of LSP resonances and flexible control of SPP transmissions. *ACS Photonics*, 2015;2(6):738–743.

[221] Ng B, Wu J, Hanham SM, et al. Spoof plasmon surfaces: a novel platform for THz sensing. *Adv Opt Mat*, 2013;1:543.

[222] Ng B, Hanham SM, Wu J, et al. Broadband terahertz sensing on spoof plasmon surfaces. *ACS Phot.*, 2014;1:1059.

[223] Cao Pan B, Liao Z, Zhao J, et al. Controlling rejections of spoof surface plasmon polaritons using metamaterial particles. *Chem Rev.*, 2008;108(2):494–521.

[224] Song K, Mazumder P. Active terahertz (THz) spoof surface plasmon polariton (SSPP) switch comprising the perfect conductor

meta-material. *2009 9th IEEE Conference on Nanotechnology (IEEE-NANO)*, 2009;56(11):2792-9.

[225] Song K, Mazumder P. Nonlinear spoof surface plasmon polariton phenomena based on conductor metamaterials. *Photonics and Nanostructures – Fundamentals and Applications*, 2012; 10(4):674-9.

[226] Wan X, Yin JY, Zhang HC, Cui TJ. Dynamic excitation of spoof surface plasmon polaritons. *Applied Physics Letters*, 2014;105(8).

[227] Sun W, He Q, Sun S, Zhou L. High-efficiency surface plasmon meta-couplers: concept and microwave-regime realizations. *Light: Science & Applications*, 2016;5(1):e16003.

[228] Sun S, He Q, Xiao S, Xu Q, Li X, Zhou L. Gradient-index meta-surfaces as a bridge linking propagating waves and surface waves. *Nature Materials*, 2012;11(5):426-31.

[229] Sun S, Yang KY, Wang CM, et al. High-efficiency broadband anomalous reflection by gradient meta-surfaces. *Nano Letters*, 2012;12 (12):6223-9.

[230] Quevedo-Teruel O, Ebrahimpouri M, Kehn MNM. Ultrawideband metasurface lenses based on off-shifted opposite layers. *IEEE Antennas and Wireless Propagation Letters*, 2016;15:484 487.

[231] Valerio G, Sipus Z, Grbic A, Quevedo-Teruel O. Accurate equivalent-circuit descriptions of thin glide-symmetric corrugated meta-surfaces. *IEEE Transactions on Antennas and Propagation*. 2017;65 (5):2695-2700.

[232] Gao F, Gao Z, Shi X, et al. Probing the limits of topological protection in a designer surface plasmon structure. *Nature Communications*, 2015;7(May):17.

[233] Khorasaninejad M., Capasso F. Metalenses: Versatile multifunctional photonic components. *Science*, 2017;358:8100.

[234] Wu H-W, Han Y-Z, Chen H-J, Zhou Y, Li X-C, Gao J, Sheng Z-Q. Physical mechanism of order between electric and magnetic dipoles in spoof plasmonic structures. *Optics Letters*, 2017; 42 (21):4521-4524.

[235] Ma Z, Hanham SM, Huidobro PA, Gong Y, Hong M, Klein N, Maier SA. Terahertz particle-in-liquid sensing with spoof surface plasmon polariton waveguides. *APL Photonics*, 2017; 11(2):116102.

Printed in the United States
By Bookmasters